THE VALUE OF THE
MOON

THE VALUE OF THE

MON

How to Explore, Live, and Prosper in Space Using the Moon's Resources

Paul D. Spudis

Smithsonian Books

Smithsonian Books
Washington, DC

Hardcover edition published in 2016.
Paperback edition published in 2022.

This book may be purchased for educational, business, or sales promotional use. For information, please write: Special Markets Department, Smithsonian Books, P. O. Box 37012, MRC 513, Washington, DC 20013

Published by Smithsonian Books
Director: Carolyn Gleason
Production Editor: Christina Wiginton
Editorial Assistant: Jaime Schwender

Edited by Gregory McNamee
Designed by Brian Barth

Library of Congress Cataloging-in-Publication Data

Names: Spudis, Paul D., author.
Title: The value of the Moon : how to explore, live, and prosper in space
     using the Moon's resources / Paul D. Spudis.
Identifiers: LCCN 2015033833| ISBN 9781588345035 | ISBN 1588345033
Subjects: LCSH: Moon–Exploration. | Outer space-Exploration. | Space flight. |
     Space industrialization.
Classification: LCC QB582.5 .S68 2016 | DDC 333.9/4-dc23 LC record available at
     http://lccn.loc.gov/2015033833

Manufactured in the USA

26 25 24 23 22   1 2 3 4 5

Paperback ISBN: 978-1-58834-564-6

CONTENTS

PREFACE

Twenty years ago, I wrote *The Once and Future Moon* (Smithsonian Institution Press, 1996). That book described the field of lunar science for the interested nontechnical reader and explained what we had learned about the processes and history of the Moon from robotic and human missions. We were acquiring some tantalizing hints that the Moon was *useful*—that it contained the material and energy resources necessary for a sustained human presence there. In the decades since then, exploration by robotic spacecraft has shown us more about the nature of these resources, confirming that the Moon is a more compelling destination than we had previously thought.

Regrettably, strategic confusion currently abounds in the American civil space program. Despite the hype and disprovable propaganda that we are preparing to conduct human missions to Mars, such an effort is far away technically, politically, and especially fiscally. A program to extend human reach beyond low Earth orbit (LEO) was arbitrarily terminated in 2010, and no rational program was offered by the administration as a replacement. Into this leadership vacuum, Congress stepped forward with a makeshift program to build a heavy lift launch vehicle (the Space Launch System) along with a human spacecraft designed for missions beyond LEO. No mission for these two items has been articulated. We will soon have some nice hardware, but no place to go.

In part, this policy chaos resulted from a misguided attempt to re-create the Apollo program. Apollo, now almost a half-century in the past, was the national effort that sent humans to the Moon. Contrary to the belief of many, the Apollo program was not about space exploration—it was about beating the Soviet Union to the Moon by landing a man there first. The entire Apollo program was a Cold War battle, and the United States won. Afterward, we stopped going to the Moon. The wartime setting of Apollo dictated that it be conducted along the lines of a wartime program: with urgency, marshalling the best technology and industrial capacity we could muster, and with cost as a secondary consideration.

Since then, we have repeatedly failed to achieve sustainable space exploration beyond LEO by trying to shoehorn it into the Apollo template. After landing American astronauts on the Moon in a highly visible and successful manner, perhaps it was natural to assume that this approach should be the configuration for future space endeavors. But after continually trying to re-create the Apollo experience by focusing on a similar human mission to Mars, with all pieces launched entirely from the Earth, we are little closer to that goal today than we were fifty years ago. The Apollo template, applied to the even greater technical challenge of a Mars mission, is enormously difficult and thus, enormously expensive, requiring tens to hundreds of billions of dollars to conduct a single mission.

A slower but affordable approach to the problem of a human Mars mission would be to gradually and incrementally increase the range of spaceflight. To do this, we would need several technical developments, including reusable vehicles based in space, staging nodes at strategic space locations, and the ability to provision ourselves for the trip from non-Earth resources, especially with high-mass, low-information density items, such as life-support consumables and rocket propellant. To our great good fortune, nature has provided us with a readily available source for this materiel—the Moon.

We can use the Moon to create new spaceflight capability. Water ice, the most useful material in space, occurs in abundance at the poles of the Moon. We can access and extract these valuable deposits because the poles also possess areas where we can generate electrical power nearly continuously. The polar "oases" of the lunar desert allow us to live on the Moon and learn how to use off-Earth material and energy resources. This effort will create a new paradigm of spaceflight: to use what is available in space instead of launching it all from the deepest gravity well in the inner solar system, the Earth's surface. Such a development will revolutionize space travel.

Of critical importance to achieving this revolution is working out how to affordably establish a presence on the Moon. We have limited time and money to spend on space. I believe that there is a path to the Moon, one that accommodates the needs of federal, international, and commercial interests, a visionary scheme that will open up the solar system to economic development.

Modern technical civilization depends on a variety of assets in space. These machines monitor our weather and environment, provide instant global communications, permit precision navigation anywhere in the world, and secure our nation and the world with strategic surveillance. Satellites are vulnerable, and a national presence in cislunar space—the space between Earth and the Moon—is essential to guarantee our continued and uninterrupted access to these assets. A robust presence by the United States in cislunar space is necessary to assure the future emergence of free markets and to promote the growth of a pluralistic, political system on the new frontier.

This book tells the story of how we once went to the Moon, what we found as a result, our various efforts to return there, and especially why and how we should go back. We go to the Moon to create new capabilities. It is the next logical step in space beyond LEO.

I thank my colleagues who critically read and reviewed all or parts of the manuscript: Sam Lawrence (Arizona State University), John Greuner (NASA–Johnson Space Center), Jack Frassanito (Frassanito and Associates, Inc.), Tony Lavoie (NASA–Marshall Space Flight Center), and Ben Bussey (Johns Hopkins University Applied Physics Laboratory, currently detailed to NASA Headquarters). Some figures were provided by Dennis Wingo (Skycorp, Inc.), Mark Robinson (Arizona State University), and Jack Frassanito. As always, my wonderful wife, Anne, is my most insightful critic, merciless editor, and best friend; I especially thank her for editing multiple versions of this manuscript and for general inspiration.

1

Luna: Earth's Companion in Space

Humans dreamed of touching the Moon for millennia. It was only within living memory that we actually left our planet and stepped upon the strange new world that lies on our celestial doorstep. Recently, an international flotilla of robotic probes mapped the properties and determined the processes of this lunar world. Amazingly, it found that the Moon contains the material and energy resources needed to establish a permanent, sustained human presence there. Water ice was found near the poles of the Moon—billions of tons of ice, trapped in its cold, dark regions. Areas close to these ice deposits are bathed in sunlight for most of the lunar year. Water and light are two resources that permit us to use the Moon to create new capabilities for spaceflight. Thus, the Moon is an object of great utility that offers us strategic and operational possibilities that other destinations in space do not.

Because the Moon is close, we can access it easily and continuously, unlike virtually any other deep space destination. The Moon's nearness means that much of the initial work of producing water and preparing the surface for habitation can be done remotely with robots under the control of human operators on Earth. Unique among space destinations, the proximity of the Moon allows us to begin its development before sending people, making the lunar surface the most inexpensive space goal beyond low Earth orbit, where significant progress can be attained early. The low gravity of the Moon (one-sixth that of Earth) enables us to use its resources to provision ourselves with the air, water, and propellant needed for the interplanetary journeys that humanity will undertake in the future.

The Moon is a small, complex satellite with a protracted and fascinating history and evolution. The early history of the solar system, a distant age when planets collided, globes melted, and crusts were formed and bombarded by impacts of leftover debris, are recorded in the rocks and soil of the Moon. The Moon has a core, a mantle, and a crust. Giant impact craters and basins have excavated thousands of cubic kilometers of rock and then crushed, melted, and reassembled it into complex forms. Internal melting generated magmas, which were released onto the surface as massive outpourings of lava, flooding large regions of the lunar surface. Following this period of violent geological events, near quiet has presided over the last billion years. The fossilized world of the Moon intrigues us, challenging our understanding of how the universe works.

All of these attributes place the Moon in the high-value column when selecting future strategic directions for humans in space. We went there half a century ago largely because a human lunar landing was a dramatic space goal achievable within a reasonable amount of time. Now, this same proximity, coupled with the Moon's intrinsic interest and resources, again makes it an attractive destination. As we consider this, it is important to know how we went before, what we learned and why the Moon is the logical next strategic goal for the American space program. I will relate the history of our efforts to return to the Moon and the multiple starts and stops of that effort. Like Sisyphus and his stone, each time we thought we were on the road back to the Moon, we seemingly rolled back to the beginning. But unlike Sisyphus, each failed attempt to restart lunar spaceflight resulted in the acquisition of new data and information that has shown us that the Moon is an even more useful and inviting destination than we had thought. It is a wandering and complex (but fascinating) story involving geopolitics, government spending, big science and technology, and national greatness.

The Moon as an Object of Wonder, Mystery, and Worship

As the largest object in our night sky, the Moon has always been an object of interest and awe. From our first gaze overhead, we have wondered about and studied it, charting its path across the heavens. Because the Moon's shape and appearance changed with regularity, it suggested to early humans that there was order in the otherwise capricious and potentially dangerous unknown world around them. The Moon allowed the earliest life on Earth to measure the passage of time, predict the seasons, and plan ahead—survival skills important to all species. Early religious speculation involved the worship of nature. The Moon's changing appearance over the course of a month, along with the passing of days and seasons, became the natural timepiece whose rhythms and cycles helped humans regulate their lives. The coincidence of the duration of the lunar cycle to human menses suggested a female presence in the heavens. In the pantheon of deities, Moon goddesses Artemis, Diana, and Selene oversaw the natural world.

Even after ancient nature worship had been largely abandoned in western culture, the Moon remained a timekeeper and an object of intrigue. Both Judaic and Muslim religious calendars are lunar-based, not solar-based. Because the lunar and solar cycles are not coincident, holidays such as Passover and Ramadan fall on different dates every year. Aside from its early, practical use as a timekeeper, the Moon also influenced culture. A full moon permitted considerable outdoor activity during preindustrial history, spawning tales and legends of werewolves and "lunacy"—the idea that a full moon (Luna) could induce unnatural and abnormal behavior and activity.[1]

We now know that Earth's Moon has been, and will remain, intimately tied to human origins, history, and development. The Moon's twenty-eight-day orbit around Earth acts as a stabilizing influence on the obliquity of Earth's spin axis, causing it to be stable for extended geological periods. Without this stabilization, rapid and chaotic changes in the orientation of its spin axis would make Earth oscillate wildly between climatic extremes, as happened on Mars. The Moon's rotation around Earth causes tides on its oceans and land, resulting in the development of periodically inundated coastal areas, sometimes below water and sometimes above it. Such terrain fluctuation is believed to have facilitated the development of land creatures, as marine species began to tolerate brief periods on dry land. Thus, because of its gravitational influence, the Moon was a major driving force in the evolution of life on Earth.

Anaxagoras (500–428 BCE) was among the first of the early Greeks philosophers to examine the Moon scientifically. He believed that the Moon did not shine from its own light, but merely reflected the light of the Sun. He also developed the first correct explanation of solar eclipses. Aristotle (384–322 BCE) believed that the Moon was a sphere, always showing the same hemisphere (the near side) to us. Aristarchus of Samos (310–230 BCE) calculated the distance between Earth and Moon at 60 Earth radii, an astonishingly good estimate (in its elliptical orbit, the Moon actually varies in distance between 57 to 64 Earth radii, or between 363,000 to 406,000 km).[2]

During the Middle Ages, leading up to the Renaissance, or roughly the fifth to the sixteenth centuries, the Moon was simply another object to astronomers, but it did play a key role in the development and evolution of modern physical science. Galileo (1564–1642), an Italian philosopher, physicist, and astronomer, not only observed the Moon with a primitive telescope but also conducted experiments on the laws of motion and was an early convert to the Copernican system of a heliocentric solar system. The recorded motions of the Moon and planets against a background of fixed stars by careful observers, such as the Danish court astronomer Tycho Brahe (1546–1601), led German scientist Johannes Kepler (1571–1630) to formulate his three laws of planetary motion. A key insight is that planets and moons orbit their primaries in elliptical paths, not circular ones, as Copernicus (1437–1543) had suggested. As the Renaissance gave way to the Age of Enlightenment, English physicist Isaac Newton (1643–1727) synthesized the observations of Tycho, and the laws of planetary motion by Kepler, into a unified theory of gravitation. Once again, the Moon played a critical role. As Newton observed an apple fall from a tree in his garden, he wondered if the force acting upon the apple was the same force that kept the Moon in its orbit around Earth. From this simple musing, he developed the laws of motion and universal gravitation, a mathematical system that explained the physical world in exquisite, clockwork detail.

Although the naked eye cannot resolve individual landforms on the Moon, patches of light and dark areas on its visible disc have been discussed since antiquity, leading to fanciful Rorschach-like interpretations, ranging from the famous "Man in the Moon" to rabbits, dogs, dragons, and a wide variety of other creatures or objects. The dark and light areas are caused by the Moon's two principal terrains: the dark, smooth *maria* (Latin for "seas") and the brighter, rougher *terra* ("land" or highlands). The association of the dark terrain with seas has a muddled history. Galileo is often credited with it, but he didn't actually equate the dark areas with water; he only suggested that some "might" be so. Using the newly invented telescope, Galileo made drawings and wrote detailed descriptions of the complex landforms that make up the lunar surface.[3] By observing the Moon during different phases and surface illuminations, he saw that its surface was not smooth, as some of the classical philosophers had surmised, but rough and jagged, consisting of towering mountains and most significantly, circular depressions in a wide variety of sizes. Even though the Moon's near side had been thoroughly mapped and remapped by astronomers over the previous two hundred years, the use of the word "crater" (from the Greek word meaning cup or bowl) to describe these holes was not used until the late eighteenth century.

With the advent of increasingly more powerful telescopes, the landscape of the lunar near side became known in much greater detail (figure 1.1). Astronomers now moved past the Moon to the more interesting stars, nebulas, and galaxies beyond. Lunar studies were left to a few diehards, mostly amateur astronomers and rogue geologists. The vast bulk of work on the Moon in the nineteenth and early twentieth centuries dealt with descriptions and studies of its surface features and history—most pressingly, the problem of the origin of craters. There were two opposing camps regarding craters. One group held that volcanic explosions and eruptions formed craters, while the other group believed that craters were made by the impact of small bodies, such as asteroids and comets.[4] This debate grew to near religious intensity, often with more heat than light being shed on the problem. The two proposed mechanisms had very different implications. The volcanic hypothesis suggested that the Moon was an active body, with internal heat and ongoing volcanism. The impact idea suggested instead that the Moon was cold and dead and might never have had any internal activity. To support their arguments, each side marshaled the best examples they could; few analogues from the study of Earth's landforms were of any help. Although Earth has many volcanoes that have been studied for years,

Figure 1.1. View of the waxing gibbous Moon generated from LRO WAC images. The dark, smooth plains (maria) are basaltic lava flows, mostly erupted before three billion years ago. The rough, heavily cratered highlands (terrae) are the remnants of the original lunar crust. Bright spots are fresh craters.

at the beginning of the twentieth century, no recognized terrestrial impact feature had as yet been described.

In 1892, Chief Geologist of the US Geological Survey Grove Karl Gilbert, intrigued by the craters of the Moon, spent many nights studying the lunar surface through a telescope at Washington's Naval Observatory. Gilbert had heard a lecture about meteorite fragments that had been collected near a feature known as Coon Butte in northern Arizona. Mineralogist Albert Foote described these iron meteorites and noted their proximity to Coon Butte, but did not go so far as to connect the two in origin. Gilbert decided to study Coon Butte as a possible impact crater. By carefully measuring the shape of the crater, he calculated the likely size of an impacting iron meteorite. He postulated that the remnants of such an object must currently exist beneath the floor of the crater and used a magnetic dip needle (designed to show variations in Earth's magnetic field) to search for what he believed should be an enormous buried iron body below the surface. But after intensive mapping failed to reveal the buried meteorite, Gilbert reluctantly (and wrongly) concluded that the Coon Butte crater must be a volcanic steam vent.[5] Today, Coon Butte is known as Meteor Crater and is considered the world's first documented meteorite impact site. How did Gilbert get its origin so wrong, especially since he had specifically tested the impact idea?

Gilbert did not understand that an impact at extremely high velocities (greater than 10 km/second) produces such enormous energies that the projectile essentially vaporizes as a point-source release of energy; left behind is a big hole with no buried iron body beneath the crater floor. An impact event is very similar to the detonation of a nuclear bomb. In fact, the formation of Meteor Crater fifty thousand years earlier by the impact of an iron meteorite must have looked very much like a nuclear explosion, complete with blinding flash and subsequent mushroom cloud. Documentation that this crater formed by impact opened the floodgates to the recognition and cataloging of dozens of impact craters on Earth (a process that continues to this day). Study of these features taught scientists to recognize the physical and chemical effects of high velocity impact, knowledge that would become critical in future interpretations of samples from the Moon and for a startling new interpretation of Earth's history as well.

The Moon as Destination: The Space Race

The idea that we might someday travel to the Moon was often the subject of imaginative fiction, but such a journey could not be seriously contemplated until Konstantin Tsiolkovsky, Hermann Oberth, and Robert Goddard had developed the basic principles of rocketry and spaceflight.[6] The technology of rockets made great strides under the impetus of war, as Germany developed the world's first intercontinental ballistic missile (ICBM), the A4 (or "V-2" as Hitler dubbed it). In the years following World War II, intensive work toward the development of larger and better ICBMs as weapons of war led to the advent of Earth-orbiting satellites (Soviet Sputnik in 1957 and American Explorer 1 in 1958) and ushered in the Space Age. War and space were tightly coupled from the beginning, since the first use envisioned for space revolved around its possible value as a battleground.

Given this background, it was inevitable that the Moon would emerge as a key object in the exploration of space. Indeed, trips to the Moon began shortly after the beginning of the Space Age with the flight of Luna 2 in 1959. This Soviet robotic probe hit the Moon after a three-day journey, making it the first man-made object to reach another extraterrestrial body. Because of the Moon's prominence in the sky and its proximity to Earth, it quickly became the focus of the first race into space between the United States and the Soviet Union. In May 1961, responding to a growing sense of geopolitical competition, President John F. Kennedy declared a national goal of a human lunar landing by the end of the decade. It was widely assumed that the USSR had accepted America's challenge and that the "Race to the Moon" was on. A series of activities in Earth orbit conducted by both nations soon followed, filling that decade with new space accomplishments, which included extravehicular activities (spacewalks), the rendezvous and docking of two orbital spacecraft, long-duration flights (up to two weeks), flights to extremely high altitudes in the hundreds of kilometers, and the mastery of complex orbital changes. All of these techniques would be needed for a human mission to the Moon.

Meanwhile, the United States launched a series of robotic spacecraft to examine and scout the Moon. These missions probed its surface, landed softly on it, examined the soil, took high-resolution images of its surface features, and prepared the way for future human missions. The Ranger (impactors), Surveyor (soft landers), and Lunar Orbiter series gave us a first-order understanding of lunar surface features, processes and history.[7] Scientists and engineers learned that the surface was dusty, yet strong enough to support the weight of a lander and astronauts. Craters covered every square millimeter of its surface, ranging in size from microscopic to enormous basins spanning thousands of kilometers. The landscape of the far side of the Moon turned out to be very different from its near side, with a near-absence of the dark, smooth maria that cover much of the Earth-facing hemisphere. Many unusual landforms of non-impact origin were found in the maria, strongly suggesting its origin as volcanic lava flows. Assuming that most craters were formed by impact, their density and distribution suggested that the Moon was an ancient world. Its surface told a story of having being exposed to space for many millions to billions of years.

The results of the Apollo missions, along with 380 kg (842 pounds) of rock and soil samples returned to Earth, largely confirmed and extended these inferences.[8] We found that the Moon is made up of some of the same rock-forming minerals widely found on Earth and that it formed almost 4.6 billion years ago, about the same time as Earth. The samples suggested that the early Moon had been nearly completely molten, covered by an "ocean" of liquid rock. After this magma solidified at 4.3 billion years, a barrage of asteroids and comets bombarded the Moon's surface for the next 400 million years, mixing-up the crust and creating a rough, heavily cratered surface. A final cataclysmic series of large impacts about 3.9 billion years ago formed the youngest basins, including the large, prominent Imbrium basin on the near side. The low areas of impact basins slowly filled with volcanic lava over the next 800 million years. For most of the last couple of billion years, the Moon has been largely inactive, with only the occasional large-body impact punctuating the slow and steady "rain" of micrometeorites that continue to grinds the surface into a fine powder.

This brief sketch of the history and evolution of the Moon describes a more complex planetary body than had been imagined before the Space Age. The Moon's scarred, ancient surface records not only its own history, but also that of impacts in the Earth-Moon system as well. Because the Moon has no atmosphere or global magnetic field, the dust grains of the lunar surface also record the particle output of the Sun for the last three billion years. With the Moon as a "witness plate" to events in this part of the universe, this geologic time capsule remains virtually untouched, waiting to be recovered and read. Although we found that the Moon is depleted in volatile elements compared to Earth, we have only explored the lunar surface with people at six sites, all relatively close to the equator and on the near side. One cannot help but wonder what possible surprises await us at the regions near the poles or on the far side.

Most people are familiar with the political and pop-culture effects the Space Race had on the world, but they are not as well versed on the profound scientific impact of the Apollo missions. For the first time, we had collected samples from another world, taken from sites of known location and geological context. We took what we learned from these physical samples and coupled it with the global data gained from the robotic precursors. Added to this knowledge was information attained from regional areas through remote sensing. Combining all of these data allowed us to reconstruct the story of the Moon with a high degree of fidelity. The most important discovery of the Apollo studies was recognition of the critical importance of the process of impact on the history and evolution of the solar system. From an elusive and questionable idea in the pre–Space Age era, the collision of solid objects became recognized as *the* dominant, fundamental process in planetary formation and evolution. Because we had learned to recognize the physical and chemical effects of hypervelocity impact through the study of the lunar samples, we soon recognized that large body impacts had occurred on Earth in the distant past. In particular, the extinction of the dinosaurs 65 million years ago was recognized to have happened simultaneously with the impact of an asteroid 10 kilometers in diameter. This observation, suggesting that impacts might cause mass extinctions of life, was soon extended to other extinction events evident in Earth's fossil record.[9] Some scientists now think that mass extinctions caused by impact may be one of the principal drivers of biological evolution. Thus, because we went to the Moon more than forty years ago, we now understand something very profound about the history of life on our home planet—an understanding that holds clues about our past and poses some sobering implications for our future.

The Moon as an Enabling Asset

For all of its impressive scientific and technical accomplishments, the Apollo program left many space advocates wanting. Because it was primarily driven by geopolitical conflict and designed to demonstrate our technical superiority, once Apollo had achieved its objective of "landing a man on the Moon and returning him safely to Earth," as President Kennedy's proclamation put it, there was no longer any reason to continue returning to the Moon or to go beyond into the solar system. Thus, the program held within itself the seeds of its own demise. The rates of expenditure acceptable during the Apollo program were simply not politically feasible for any follow-on space program.[10] So the decision was made to make an attempt to lower the cost of spaceflight via a reusable space shuttle. While this effort did not succeed in lowering costs, the development of the shuttle led to some significant and unique capabilities. More importantly, it pointed the way toward an alternative architectural template for spaceflight, one in which small pieces, incrementally launched and then assembled in space and operated as a large system of systems, might multiply spaceflight capabilities carried out over a longer, more sustainable period of time. This template of operations reached its acme with the completion of the International Space Station (ISS).

As for missions to the Moon, there was only silence and isolation. Several attempts to fly an unmanned orbital mission to obtain additional global remote sensing data (which would permit better interpretation of the superb Apollo sample database) were unsuccessful. With the focus of the human program centered on the space shuttle and the subsequent building of a space station in low Earth orbit, little interest in additional lunar exploration was evident. Then, in the mid-1980s, a confluence of events occurred to focus attention once again on the Moon, an interest that continues to the present. First came the realization that after the building of the space station, an orbital transfer vehicle designed to reach high orbits, such as geosynchronous (~36,000 km or 22,000 miles high), was the obvious next step. A vehicle that can reach geosynchronous Earth orbit (GEO) can also reach the Moon. Thus, a series of studies focused on the possibility of lunar return, with an emphasis on longer, more permanent stays on the surface.

The idea that we might want to remain on the Moon for longer periods of time inevitably led to the concept of obtaining some supplies locally, from the materials and energy found and available on the Moon. This concept, called in situ resource utilization (ISRU), is an essential skill for humans to master if we are to be significantly and permanently present in space and on other worlds.[11] That realization led to a renewed interest in getting additional lunar data—most especially, data for the unique local environment found at the Moon's polar regions. Because the spin axis of the Moon is nearly perpendicular to the ecliptic plane (figure 1.2), the Sun is always on the horizon at the poles. Some areas are in permanent darkness and hence, very cold. It was recognized that these "cold traps" might contain deposits of ice, along with other volatile substances deposited over geological time as water-bearing comets and asteroids collided with the Moon's surface. Additionally, other areas near the poles might be bathed in permanent sunlight. This near-continuous energy source allows for the generation of electrical power during the long, two-week lunar night. At the time, we did not know the details of these hypothesized properties or even if they actually existed. However, over the past twenty years, a number of lunar robotic missions have revolutionized our knowledge of the Moon, and in particular the environment and deposits of the poles.

In 1994, the Department of Defense Clementine mission mapped the mineralogy and topography of the entire Moon from orbit. An improvised experiment on this flight used the spacecraft transmitter as a radio source to illuminate dark areas within craters near the poles. Analysis of radio echoes from the south pole suggested the presence of water ice in the crater

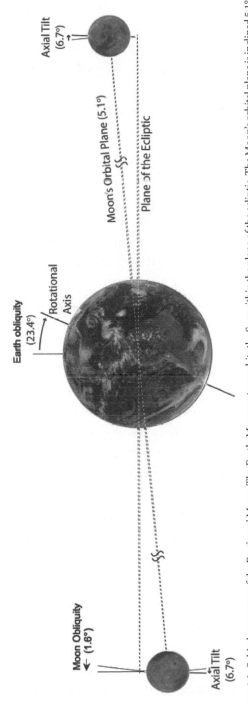

Figure 1.2. Orbital geometry of the Earth and Moon. The Earth-Moon system orbits the Sun within the plane of the ecliptic. The Moon's orbital plane is inclined 5.1° from the ecliptic, and the Moon's spin axis is tilted 6.7°. This results in a nearly perpendicular orientation of the Moon's spin axis to the ecliptic (called obliquity) of 1.6°. This is in contrast the Earth's obliquity of 23.4°.

Shackleton. This discovery was confirmed a few years later by the Lunar Prospector spacecraft, which found enhanced amounts of hydrogen at both poles. These discoveries stunned the lunar science community, since earlier results from the study of the Apollo samples had suggested that the Moon was bone-dry and always had been. Now, that concept—and our understanding of the Moon and its history—had to be reevaluated. Over the next few years, additional results from sample studies, remote sensing, and theoretical modeling culminated in the unequivocal detection of water vapor and ice during the impact of the LCROSS spacecraft, thus demonstrating beyond any doubt that significant deposits of water ice are present at both lunar poles. Conservative estimates of the amount of water ice run between several hundred million to more than a billion tons at each pole. Additionally, we have found that small areas near both poles are illuminated by the Sun for extended periods of time, some for more than nine-tenths of the year. All of this new lunar data has countries around the world planning ways to access the energy and resource bonanza at the poles of the Moon, available to those who arrive first.

Materials and energy are available on the Moon, two critical requirements for extended human presence. Water, in its decomposed form of hydrogen and oxygen, not only supports human life but is also the most powerful chemical rocket propellant known. Near-permanent solar energy is available proximate to the water-rich cold traps at the poles. The previously misleading image of the Moon as a barren, useless wilderness (as painted by Apollo results) has given way to a richer, more inviting, useful persona. The world now knows that the Moon is not simply another destination in space—but that it is an important enabling asset for spaceflight. Our current understanding of the Moon is vastly different from those early humans who first gazed up, grateful that they had the Moon to mark their calendars and chart the seasons. We now understand that the Moon is a world in its own right, an object located in our cosmic backyard whose resources we can access and use to travel throughout the solar system.

Our Future on Luna

Space engineer and visionary Krafft Ehricke once said, "If God had intended man to be a space faring species, He would have given him a

Moon."[12] This tongue-in-cheek statement is even more applicable today than when Ehricke first said it more than thirty years ago.

Why is the Moon a destination for humanity? Because it can be used to open up the frontier of space through the development of its material and energy resources. By harvesting the water ice and solar power available at the poles of the Moon, we create the ability for long-term human presence on the Moon and in near-Earth space. Water can fuel a permanent, reusable space transportation system that can access not only the lunar surface but also every other point between Earth and Moon. This zone, called cislunar space, is where 95 percent of our satellite assets reside. The ability to reach these places with people and machines will allow us to build space systems of extraordinary power and capability. Moreover, such a system can also take us to the planets beyond Earth and its Moon.[13]

We can use the Moon to learn how to live and work effectively and productively on another world. This goal requires us to learn how to build protective shelters, safe from the thermal and radiation extremes of deep space. To provision ourselves, we must learn how to extract our supplies from local resources, including life support consumables, and learn how to build infrastructure using local resources for construction materials. Once established on the lunar surface, we will use these new capabilities to explore our nearest neighbor in space as well as to build a "transcontinental railroad" in cislunar space and establish a permanent beachhead off Earth. On the Moon, we will learn how to explore a planet using the optimum combination of people and robots, each doing the tasks at which they uniquely excel. Finally, we will reveal and decipher the record of planetary and solar system evolution recorded in the rocks of the Moon. Some mysteries uncovered by the Apollo explorations revolutionized Earth science. Additional exploration will reveal even more startling secrets and continue to revolutionize our understanding of the world and universe around us.

Why is it important for the United States to make the Moon a high priority goal? Because the United States is not the only nation interested in it. This coin of international interest has two sides. On the positive side, our partners in current space endeavors, such as the International Space Station, have expressed great interest in human missions to the Moon. Some have begun the process of gathering detailed information from precursor robotic missions to enable future human missions to the Moon. How can we proclaim world leadership in space if we ignore a prominent destination that so many other nations are anxious to visit and exploit? Nations such as China have plans to explore and use the Moon with both robotic machines and with people. While their lunar intentions appear benign at present, they are developing capabilities now that could pose a threat to the security of this nation and other countries in the near future. Thus, there is a strategic dimension to American lunar presence. It is vital to the security and economic health of the community of nations that future societies in space develop according to pluralistic, democratic principles and that commerce is open to free markets, with respect for property rights and contract law. Although American presence in cislunar space does not guarantee such an outcome, our absence from this theater could well result in the reverse.

What makes the Moon both important and unique? It is close, interesting, and useful. The close proximity of the Moon to Earth means that we can always and easily access it, unlike the limited and infrequent launch windows to all other planetary targets. This nearness means that much of the early preparatory work at the Moon can be done by robots on the lunar surface, as directed from Earth. Thus, the first humans to return to the Moon can arrive at a fully functional, turnkey lunar outpost, assembled in advance by these teleoperated robots. Interest in the Moon derives from its role as a small planet of complex and interesting process and evolution. The Moon's environment permits unique and specialized scientific and engineering experiments to be conducted—studies not possible anywhere else in the solar system. We will find the answers to questions surrounding our moon's complexity and gain a fuller understanding of our home planet's early evolution. The utility of the Moon lies in its material and energy resources, the access to which will allow us to acquire the knowhow and means for humanity to plant its first foothold on another world.

2

The Moon Conquered— and Abandoned

How is it that America went to the Moon, going from nearly zero space capability to the lunar surface in less than a decade, and then rapidly left it? Why have we not been back since? Within that tale are important lessons, some never fully absorbed by either historians or our national leadership. Millennia before we achieved it, humans dreamed about going to the Moon. The actual circumstances of our journey had not been imagined by science fiction authors and as a result, virtually all science fiction dealing with lunar travel made the first landing the beginning (not the end) of humanity's movement into space. Now, it remains to be seen whether our first steps on the Moon really was an ending, or merely the prelude to a delayed golden age of spaceflight.

Journeys to the Moon in Fiction and Fact

The idea that some day we would be able to journey to the Moon is very old, conceivably going back as far as the early cave dwellers. The first literary description of trips to the Moon, Sun, and other heavenly destinations was likely the work of Lucian of Samosota (125–180 CE). Johannes Kepler, the discoverer of the laws of planetary motion, wrote *Somnium* ("Dream," published posthumously by his son in 1634). In his novel, Kepler describes a trip to the Moon and the view of Earth and the solar system from its surface. English clergyman John Wilkins wrote several books about trips to the Moon, the most famous being *The Discovery of a World in the Moone* (1638). In it, he outlined the idea that someday people might inhabit the Moon. Included in Wilkins's work were exotic and infeasible techniques on how to get there, such as transport by angels or with the help of harnessed fowl.

During the Industrial Age, authors of classic science fiction took more reasonable (if still fanciful) approaches to the problem of lunar flight. In Jules Verne's *From the Earth to the Moon* (1865), voyagers were shot from a huge cannon (the *Columbiad*) in order to reach escape velocity and land on the Moon. Verne skipped over how the acceleration from a cannon shot would create enormous g-forces that would kill his crew; he also misunderstood the nature of weightlessness by having his passengers experience it only when his moonship crossed the gravitational spheres of influence between Earth and Moon. Konstantin Tsiolkovsky, the inventor of astronautics and the first to derive the rocket equation, was inspired by Jules Verne and penned his own novel, *On the Moon* (1893). The character in Tsiolkovsky's story wakes up on the Moon and experiences the unusual effects of being on an alien world. Curiously, considering his contributions to rocket science, Tsiolkovsky does not record the details of the trip from Earth. The approach of H. G. Wells was even more fantastic: A special substance called Cavorite (named for its inventor, a character in Wells's 1901 novel *First Men in the Moon*) cuts off the force of gravity, allowing his sphere to effortlessly travel to the Moon. Once there, the voyagers find large insectlike creatures that live below the lunar surface.

The Moon was lifted out of the realm of fiction and fantasy and put back into the domain of science with the advent of modern rocketry (an outgrowth of the Second World War). Starting with a few eccentrics, the Moon once again became a topic for scientific inquiry. Ralph Baldwin, an astronomy student at the University of Chicago, after noticing the spectacular series of telescopic photographs on display in the lobby of the Adler

Planetarium, began cogitating about the origin of craters, basins, and the evolution of the lunar surface. He wrote down thoughts for a couple of articles before being pressed into war service, where he helped develop the proximity fuse. After the war, Baldwin collected his lunar ideas into a book, *The Face of the Moon* (1949).[1] This pre–Space Age synthesis was a fairly complete and accurate account of the Moon's processes and history—how the craters and basins were formed by impact, that the dark smooth maria were volcanic lava flows (Baldwin correctly identified them as basalt), and that the Moon's surface was very old compared to that of Earth's. Baldwin's study of the Moon continued throughout his life, and he lived to see virtually all of his surmises validated through the exploration of the Moon by the Apollo missions.

Shortly after this work appeared, the noted science fiction author Arthur C. Clarke published *The Exploration of Space* in 1951.[2] Clarke outlined an expansive vision of the future, including rockets into Earth orbit, trips to the Moon, and voyages to the planets. Interestingly, he made some careful and prescient observations about the issues of landing and sustaining a permanent human presence on the Moon. Clarke considered the Moon an essential way station on the road to the planets. Here humans would learn the techniques of exploring and living on an alien world. Clarke specifically recognized that using the mineral resources of the Moon to support human presence and create new capabilities was essential. He pointed out that, at least in the early phases of operation, centralizing operations at a single site on the Moon would permit concentration of resources to maximize capabilities quickly. Thus, Clarke advocated building a base, not multiple sortie missions to many different locations. After the establishment of a presence at a base, we would be able to explore the entire Moon at our leisure.

Accounts hold that Nobel Prize–winning chemist Harold Urey became engrossed by *The Face of the Moon* when, by chance, he picked up the book at a party. Baldwin's description of the lunar landscape and the impact origin of its craters convinced Urey that the primitive, ancient Moon held secrets to the origin of the solar system. He went on to lead an effort that applied the basic principles of chemistry and physics to the origin and evolution of the Moon and planets.[3] Another astronomer, Gerard Kuiper, held the "heretical" view that the Moon and the planets were worthy objects for observation and scientific study. For further study and mapping, he collected the best telescopic images of the Moon at the Lunar and Planetary Laboratory that he established in 1960 at the University of Arizona in Tucson. Geologist Eugene Shoemaker, who was mapping uranium deposits in northern Arizona for the US Geological Survey in the 1950s, decided to reexamine the geology of Coon Butte, the feature dismissed by G. K. Gilbert as not being an impact structure sixty years earlier. Using the geology of the crater to unravel the mechanics of hypervelocity impact, including the discovery of forms of silica created only under extremely high pressures, Shoemaker decided that Coon Butte was an impact crater. It has been known as Meteor Crater ever since.[4]

But Gene Shoemaker did more than just document the impact origin of Meteor Crater. In 1960, he made the first geological map of the lunar surface, showing the basic sequence of events that had occurred there. In brief, this technique involves using overlap and superposition relations to classify laterally continuous rock units, including sheets of crater ejecta and lava flows. These properties can be determined directly from visual observations and photographs. Shoemaker mapped the region near the crater Copernicus on the near side, working out the basic framework of lunar stratigraphy—that is, the sequence of layered rocks.[5] He then used this information to estimate the time correlation between events on the Moon and those on Earth, concluding that the Moon preserved an ancient surficial record, which holds part of the early geological story missing from the eroded and dynamic surface of Earth.

These scientists and their research, each in their own way, made the study of the Moon and its processes scientifically respectable. After the launch of the first Earth-orbiting satellite Sputnik 1 in October 1957, it was reasonable to imagine that spacecraft might be sent to the Moon. Soon, observations of the Moon's surface through telescopes, the mapping of terrestrial impact craters, and compositional studies of rocks from terrestrial impact craters and meteorites (rocks from space) became part of cutting-edge lunar science. A gradual but perceptible momentum began to formulate a conceptual model that would allow us to explore the Moon effectively and give us an understanding of its history. Some dreamed that they might live to see people travel to the Moon in their lifetimes (and Shoemaker planned on being one of them). Shoemaker's dream would come true in part, but under circumstances that no one foresaw.

Tho Apollo Program

In a series of articles published in *Collier's* in the 1950s, rocket scientist Wernher von Braun outlined a plan to send people to the Moon and to Mars.[6] Accompanied with colorful illustrations by space artists such as Chesley Bonestell, von Braun's articles caught the imagination of the public, including a very imaginative Walt Disney, who went on to feature von Braun's ideas in a series of programs as part of his new television series *Disneyland* (1954). Viewers were treated to a four-program series outlining the basic von Braun architecture: rocket to Earth orbit, space station, Moon tug, and human Mars spacecraft. This steppingstone approach made both logical and programmatic sense. Each piece enabled and supported the next step into space. Although some technical details in von Braun's plan were out of date before they were realized—for example, von Braun had electrical power in space generated by solar thermal power alternately vaporizing and condensing mercury to drive turbines, a technology made obsolete with the advent of solar photovoltaic cells—major parts of his scheme enabled the establishment of a robust and permanent spacefaring system.

However, international events soon intervened on von Braun's orderly approach. The advent of the Apollo program altered what was to have been a logical, incremental, and thoughtful space plan into a race once competition with the Soviet Union became our overriding concern. The slow approach had to be accelerated once President Kennedy committed the nation to a decadal deadline. Under ordinary technical development, each piece would be designed, built, flown, and modified according to its performance. But with scheduling pressure designed to beat the Soviets to the Moon, a much faster approach was required. This caused von Braun and others at the newly created National Aeronautics and Space Administration (NASA) to reexamine the problem of sending people to the Moon. Did we really need a space station first? Or was it possible to build a launch vehicle big enough to send an entire expedition to the Moon in one fell swoop?

Although the space agency had already begun planning for the development of a new super heavy lift rocket and had done some preliminary studies of manned missions to the Moon, the announcement of a lunar landing goal by President John F. Kennedy in May 1961 shocked many at NASA. It was one thing to daydream about sending people into deep space and to the Moon, but quite another to actually be given the task to do so—and then bring them back safely to Earth, a stipulation of Kennedy's declaration. When the commitment to go to the Moon was made, the total manned spaceflight experience of the United States consisted of Alan Shepard's fifteen-minute-long suborbital hop. A lunar voyage would require the mastery of a variety of complex spacefaring skills, including precision navigation and maneuvers necessary to change orbit in flight.

The design or "architecture" for a manned lunar mission was debated extensively before the "mode decision." Initial plans called for either a direct ascent to the lunar surface or a rendezvous of two launched spacecraft in Earth orbit. Both approaches called for the development of a "super" heavy lift launch vehicle, Nova, a rocket capable of launching up to 180 metric tons to low Earth orbit.[7] John Houbolt, an engineer at Langley, advocated instead for something called lunar orbit rendezvous.[8] This called for a small vehicle that would land on the lunar surface, then return to rendezvous with the Apollo spacecraft that had remained in orbit around the Moon. Although this mission profile was thought to be very risky (a rendezvous had never been accomplished in space, let alone one involving two separate spacecraft orbiting the Moon), it did enable the voyage to be launched "all up" on a single heavy lift rocket. This design became the Saturn V, a rocket capable of launching 127 metric tons to low Earth orbit.

With the principal design features of Apollo outlined, the American space program next undertook a series of manned and unmanned missions in preparation for a lunar landing. While human missions practiced specific techniques (including rendezvous and docking), robotic missions gathered information about the Moon's surface conditions and environment and sought to identify a smooth, safe landing site. In preparation for the Moon, we flew six single-man Mercury missions, ten two-man Gemini missions, and four three man Apollo rehearsal flights. There were thirteen successful robotic precursor missions to the Moon: three hard-landers, five soft-landers, and five orbiters. All this occurred within the eight years between Kennedy's speech and the landing of Apollo 11, a span of time that included the assassination of President Kennedy on November 22, 1963, and a twenty-two-month stand-down after the tragic fire on Apollo 1 of January 27, 1967, which killed astronauts Virgil "Gus" Grissom, Ed White, and Roger Chaffee.

The Apollo spacecraft was extensively redesigned and modified after the Apollo 1 fire. Following a highly successful checkout of the newly refurbished Command-Service Modules in Earth orbit on Apollo 7, an eleven-day mission in October 1968, the planned journey of Apollo 8 to orbit the Moon seemed to be a bold, even reckless move. After all, a spacecraft sent to the Moon without any rescue capability using a lunar module (LM) could have ended in tragedy, as was demonstrated a few years later during the Apollo 13 mission. We now know that there was a reason, one that was withheld from the public at the time, for sending Apollo 8 on a lunar journey. The CIA had intelligence that the Soviets were planning a manned flight around the Moon by the end of 1968. They had just completed a circumlunar mission with their unmanned Zond 8, which demonstrated that the pieces for such a flight were ready. It was believed, probably correctly, that if the Soviets were able to pull this off, they would then claim to have won the Moon race, making an actual lunar landing irrelevant.[9] This possibility lent urgency to flying a manned lunar mission as soon as possible, even one that simply orbited, rather than landed on, the Moon. So, just before Christmas in 1968, Apollo 8 orbited the Moon carrying

Frank Borman, Jim Lovell, and Bill Anders. Although it was not evident at the time, the flight of Apollo 8 effectively won the Moon race for the United States.

The next two missions qualified the Apollo lunar module in Earth orbit and in lunar orbit. Then, on July 20, 1969, Apollo 11 landed two men on the Moon. There were a few heart-stopping moments when the ship's computer sent the Apollo 11 LM *Eagle* toward a large, block-strewn lunar crater, but astronauts Neil Armstrong and Buzz Aldrin successfully overrode the automatic system and landed safely. Initial concerns about possible dangerous surface conditions were soon dispelled as the crew conducted a successful 2.5-hour exploration of their immediate landing site. They collected rock and soil samples, laid out experiments, and verified that the surface was both strong enough to support the considerable mass of the LM as well as other equipment. The world watched as they demonstrated what it was like to move around on the Moon in one-sixth the gravity of Earth. Armstrong made an unreported traverse to a blocky crater that he had flown over during his landing approach and observed the bedrock in the crater floor. Twenty-two hours later, the two-man crew blasted off the Moon's surface to rendezvous with Mike Collins, orbiting the Moon in the Command Module *Columbia*. With the crew's safe return to Earth a few days later, Kennedy's daring challenge to America eight years earlier was fulfilled.

Next came the question of what to do with the remaining Saturn rockets and Apollo hardware, the surplus equipment that had been procured in the event that more than a single attempt would be needed to successfully land on the Moon. Initially, Apollo engineers planned for more lunar missions and ultimately a human mission to Mars. However, it soon became apparent that the national will was not inclined toward additional human exploration beyond the Moon—or even to it. An ambitious program to push the boundaries of human reach into space was shelved.[10] Apollo continued for a few more flights, but lunar bases and Mars missions were not in the cards. Our focus shifted to completing the Apollo program, then to developing a reusable transport-to-orbit vehicle for people and cargo—the flight program that designed, built, and operated the space shuttle.

Despite the political decision to abandon the capabilities of the Apollo-Saturn system, NASA was able to wrangle permission to fly out part of the remaining original plan for Apollo lunar exploration. Several interesting landing sites were selected for these missions, most of which had advanced capabilities and tools for exploration. Even with some notable mission problems, flight and surface operations steadily improved. Despite being struck twice by lightning during liftoff, Apollo 12 successfully landed on the Moon in November 1969. This mission validated the technique of pin-point landing by setting the LM *Intrepid* down within a hundred meters of Surveyor 3, a previously landed robotic probe. This technique allowed us to safely land at future sites of high scientific (but dangerous operational) interest. After the disaster and near-loss of the Apollo 13 mission, following the explosion of an oxygen tank in its Service Module, which cancelled its landing on the Moon, the Apollo 14 crew traveled to the highlands of Fra Mauro in early 1971. Here it was expected that they would find material thrown out from the largest, youngest impact basin, the Imbrium Basin. From this site, the astronauts returned complex, multigenerational frag-mental rocks called breccias, parts of which dated to the earliest era of lunar history.

On the final three Apollo missions (15, 16, and 17), the astronauts spent longer times on the surface and possessed greater capability for explora-tion.[11] The first three lunar landings had no surface transport, so the crews had to stay within a few hundred meters of their LM and could not remain outside the spacecraft for more than four to five hours at a time. The next three landings used more capable spacecraft and each mission carried a surface rover—a small electric cart strapped to the outside of the LM. Once they were on the surface, the cart was taken off, unfolded, and then driven by the crew to locales several kilometers away from the landing points. In addition, a redesigned spacesuit allowed moonwalks of up to eight hours duration. Consequently, an extraordinary amount of high-quality explo-ration was conducted on these latter missions. Each subsequent mission improved upon the total distance traveled, the amount of samples collected, the experiments performed and the data gathered. These were the "J-missions," and because of them, the Apollo program wrote great chapters in the history of human exploration.

Apollo 15, the fourth manned lunar landing, was sent to the rim of the Imbrium basin, at the base of an enormous mountain range called the Montes Apenninus (Figure 2.1). The mission occurred between July 26 and August 7, 1971. Astronauts Dave Scott and Jim Irwin spent three days exploring the mountains and the mare plain that surrounded them. The landing site was also near Rima Hadley, a winding, sinuous canyon believed to have been carved by flowing lava. With the Apollo 15 astronauts well trained in the sciences, especially field geology, this mission demonstrated a new and growing sophistication in lunar exploration. The astronauts found and returned a fragment of the original lunar crust, the "Genesis Rock," and

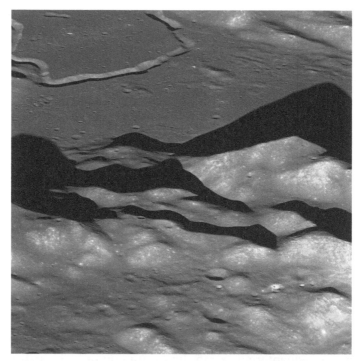

Figure 2.1. Oblique view of the Hadley-Apennine region, landing site of the Apollo 15 mission in 1971. The site is at top left; the sinuous channel near the top is Hadley Rille, a channel carved by flowing lava.

an unusual emerald green glass, created by volcanic fire fountains erupting more than three billion years ago. They also used a power drill to recover a core of the upper three-meters of the regolith at the landing site.

Continuing in this mode of surface exploration, the Apollo 16 mission visited the central lunar highlands in April 1972. Veteran astronaut John Young (Figure 2.2) and rookie Charlie Duke explored two large impact craters situated in the mountainous Descartes highlands region, northwest of Mare Nectaris. Against expectations of finding volcanic ash flows, the crew discovered instead that the highlands are made up of ancient rock debris, shattered and broken by eons of cataclysmic, large-scale impacts.

Figure 2.2. Apollo 16 Commander John Young explores the geology of the Descartes Highlands landing site. The samples and data returned from the Apollo missions are the principal sources of detailed information on lunar history and processes.

Although the astronauts did not find the expected volcanic rocks, the results of this mission led us to a better understanding of the importance of impact in the creation of the lunar highlands. The breccias found at the Descartes site may have come from one of the large multi-tiring impact basins, such as the magnificent Orientale basin on the Moon's western edge.

Illuminating the Florida landscape with a brief false dawn, the last Apollo mission to the Moon, Apollo 17, was the first night launch of the program. This mission is renowned for the first flight of a professional scientist to the Moon, LM pilot and geologist Jack Schmitt. He and Gene Cernan spent three days exploring the Moon's valley of Taurus-Littrow on the eastern edge of Mare Serenitatis, in a low region of smooth mare lavas situated between two enormous basin massifs. They found ancient crustal rocks, old mare lavas, and most spectacularly, orange soil (fine orange and black glass particles, pieces of lunar ash erupted from a lava fire fountain over 3.5 billion years ago). The magnificent scenery of the landing site and the abundant scientific return from the Apollo 17 mission was a fitting conclusion to the Apollo program. The 380 kilograms of lunar rock and soil in the sample vaults at NASA's Johnson Space Center in Houston are a lasting scientific legacy and testament to the achievement of the Apollo program.

Post-Apollo Legacies

Then it was over. When the last crew departed the Moon on December 14, 1972, no one knew when, or if, humans would return. Forty years on, Apollo 17 Mission Commander Gene Cernan remarked that he never would have imagined we would still be looking forward to man's return to the Moon. Will we go back before the fifty-year milestone, or was it all just a big, one-time stunt? Did Apollo give us something of lasting value? What is the legacy of the Apollo program? And what does it have to tell us about our future in space and about America as a spacefaring nation?

The scientific legacy of the Apollo program is remarkable. The lunar samples have been studied more intensely than almost any other collection of material in the history of science, with some rocks taken apart atom by atom. These small pieces of another world have a scientific value not present in meteorites because we know exactly where they come from on the Moon, and that information allows us to interpret their history in a broader, regional-to-global context. By reading the historical record found in the lunar samples, we have reconstructed the story of an ancient world, one where entire globes of liquid rock crystallized to form the crust and mantle of the Moon. This episode was followed by an intense bombardment—giant impacts that formed the large overlapping craters and basins of the highland surface. Remelting of the deep interior created magmas that forced their way up through the mantle and crust, erupting onto the surface as extrusive lava flows. In some cases, the amount of volatiles dissolved in the liquid rock were so great that sprays of molten rock shot into space, then quickly cooled into the fine glass spheres that make up the dark ash deposits of the Moon.

The impact bombardment of the Moon was very intense early in its history, but tapered off drastically around 3.9 billion years ago and continues at a very low intensity to this day. Most of the debris hitting the Moon now consists of micrometeorites that constantly "rain" down upon the surface. This long-term process has ground the lunar surface into a fine powder. When these tiny particles hit previously made soil, some of the soil grains are fused into a melted mixture of glass and mineral fragments called agglutinates. Because the Moon is exposed directly to space and possesses no global magnetic field, its surface is implanted with solar wind gases— particles emitted by the Sun and galactic cosmic rays, mostly protons, or hydrogen ions, that induce radiation damage in the Moon's dust grains. Thus, although the geological evolution of the Moon continues to this day, surface erosion happens at an extremely slow pace, about a centimeter every twenty million years.

From study of the lunar samples, we now understand the telltale signs of hypervelocity impact, which include both chemical and physical effects. Chemically, we can detect the small addition (on the order of a couple percent) of meteoritic debris in the lunar soil in the form of excess amounts of siderophile (iron-loving) elements such as nickel and iridium. Physically, in addition to the shock-melted glass agglutinates (mentioned above), we also see shock damage to the mineral grains of lunar rocks. The common mineral plagioclase is often turned into glass called maskelynite by impact shock, a transition that occurs without melting. Other features, diagnostic of the passage of a shock wave, include cracks, mosaicism (shattered grains that arrange themselves into geometric patterns), and lines of planar deformation. All of these chemical and physical features are found in and around terrestrial impact craters. Their occurrence in lunar samples verifies that the craters of the Moon are of impact origin.

An interesting and important consequence of this science only became apparent several years after the end of the Apollo lunar missions. Working with marine sedimentary rocks in Italy, geologist Walter Alvarez wanted to know their rates of deposition. His father, physicist Luis Alvarez, suggested that he measure the concentration of the element iridium in the rocks. Iridium is a rare element in Earth's crust, but it is more abundant in meteorites. His thought was that meteoritic debris constantly rains onto Earth at a known rate and that it could serve as a clock for measuring the rates of carbonate sedimentation on the sea floor.

When the iridium was measured, surprisingly large amounts were found in the clay layer that marks the end of the Cretaceous Era. This Cretaceous-Tertiary (KT) boundary is demarked by a thin clay layer all over the world and is the youngest horizon below which dinosaur fossils are found. This discovery advanced the idea that a massive meteorite impact sixty-five million years ago was responsible for the extinction of the dinosaurs and several other fossil families.[12] Later, small grains of shock-deformed quartz were found within the KT clay layer, supporting the idea of a large body impact occurring at that time. Subsequently, it was found that in some cases, similar boundary layers that marked mass extinctions in the geological record also contained evidence for large body impacts.

This connection was made by recognizing the critical defining evidence for hypervelocity impact, a process learned from the collection and study of the Apollo lunar samples. It was often claimed in the immediate post-Apollo period that the Moon effort had all been for naught, scientifically. It was thought that we got some rocks and some ages for a few ancient events in the history of the Moon—but so what? That "so what" is now recognized as a revolutionary paradigm shift in our understanding of the significance of impact in Earth history. We now view the process of the evolution of life on Earth from a new and unexpected perspective. Because we journeyed to the Moon, a new concept of how life may evolve was discovered.

Most of what we now know about the timeline for the origin and evolution of our solar system is tied to facts obtained from our study of the Moon. Results from Apollo scientific work carry over into all of planetary science. The concept of a late heavy bombardment (that is, the apparent increase in the cratering rate between 4.0 and 3.8 billion years ago) and estimates of the timescales upon which events on Mars, Mercury, and other objects have occurred are all reliant on the dates provided by the Apollo lunar samples. Additionally, when requesting lunar samples, investigators had to show they could make their analyses on the smallest amount of material possible. This stringent requirement forced scientists to develop techniques capable of analyzing extremely small amounts of material. This work succeeded to such an extent that fully valid analyses are now done on mere specks of dust. In addition, because some samples were very complex, such as the impact breccias of the highlands, new techniques were developed that can reveal the interior structure of such aggregate rocks using X-ray tomography, a method similar to magnetic resonance imaging (MRI) of the interior of the human body.

The political legacy of the Apollo program was no less significant than its scientific one. Despite subsequent claims to the contrary, it is now clear that the Soviets had accepted Kennedy's challenge of sending a human to the Moon and returning him safely within the decade.[13] The race to prove the superiority of an ideology had been joined. Each country needed to harness greater technology and science in order to win. This breathless competition in space was conducted with a seriousness that we can scarcely credit these days, with each new "first" being heralded as *the key* to space success, and, by inference, global domination. The Soviets orbited the first satellite, the first man, the first woman, and were first to hit the Moon with a man-made object. They orbited the first multiple-man crew, and in 1965, one of their cosmonauts, Aleksei Leonov, made the first "walk in space" when he floated outside his space-craft. America stumbled at first but rapidly caught up, matching most Soviet achievements. Soon we began making our own space firsts—the first rendezvous and docking in orbit, the first long-duration space walks, and the first successful flight of the giant Saturn V booster. But everyone knew the true high-stakes measure of success was to be the first to reach the Moon with people.

While Americans were enjoying the trill of victory with the epic flight of Apollo 11, the Soviets were having some difficulties. The Soviet Moon rocket, the gigantic N-1, a vehicle comparable in size to the American Saturn V, failed all four times it was launched.[14] These disasters, kept secret for twenty-five years, sealed the fate of the Soviet Moon program. Without an operational heavy lift booster to deliver their spacecraft, no Soviet lunar mission was possible. American democracy and free-market capitalism had outmatched the USSR and won the Moon.

In programs of vast technical scope, particularly those requiring the practical application of high technology such as high-speed computing to very complex problems, Americans had shown the world both dogged determination and technical prowess for accomplishing whatever they set as their goal. The Soviets viewed America as having achieved through a combination of great wealth, technical skill, and resolute determination an extremely difficult technological goal, one that they themselves had vigorously attempted but had failed to achieve. America's victory of getting to the Moon first and exploring its surface carried over, later figuring in a more serious conflict between the United States and the Soviet Union.

In 1983, President Ronald W. Reagan called upon the scientific and technical community of the United States and the free world to develop a system to defend the country against ballistic missiles, one that would make

America and other nations free from the fear of nuclear annihilation. This program, the Strategic Defense Initiative (SDI), was specifically conceived to counter the prevailing strategic doctrine of mutually assured destruction (MAD), whereby a nation would never start a nuclear war because it would fear its own destruction by retaliatory strikes. The price of peace in a MAD scenario was to live in a permanent state of fear. The promise of SDI was to eliminate that fear by having a system designed to defend countries from nuclear missile attack.

The Strategic Defense Initiative was roundly criticized and belittled by many in the West who considered it "destabilizing." Numerous scientists, including those who had previously done weapons work, criticized it as "unachievable." Arms control specialists decried "Star Wars," as they called it, as provocative and an escalation of the nuclear arms race. Reagan did not retreat and insisted that SDI proceed. The number one foreign policy objective of the Soviet Union in the last years of its existence was the elimination of SDI. The famous Reykjavik Summit of 1986 collapsed on this very point when Reagan would not agree to crippling restrictions on SDI deployment in exchange for massive cuts in ballistic missiles by Gorbachev and the Soviets.[15]

If the bulk of academic and diplomatic opinion was so averse to SDI and the very idea of missile defense was so "unworkable," why then did the Soviet Union fight so long and fiercely against it? Clearly, it was because the leaders of the Soviet Union were convinced that SDI *would* work—that the United States always achieved its stated goals. Because America had attempted and successfully achieved the difficult and demanding technical goal of reaching the Moon, it made any similar goal that we set out to do seem equally achievable. Moreover, this was a goal that the Soviets themselves had attempted and failed to achieve. With the specter of the American Apollo victory fresh in their minds, the Soviets had no choice but to spend whatever resources were necessary to compete with Reagan's SDI program. In the end, they went bankrupt, and their communist economy collapsed—a very real and practical consequence of America's successful Apollo program.[16]

Begun as a strategic Cold War gambit under President Kennedy, Apollo and the race to the Moon demonstrated to the world the superiority of America's free and democratic way of life over that of our communist adversaries, an achievement still not fully appreciated today. America had achieved technical credibility from the amazing success of the Apollo program. When President Reagan announced SDI twenty years later, the Soviets were against it, not because it was destabilizing and provocative, but because they believed we would succeed. That success would render their vast military machine, assembled at great cost to their people and economy, obsolete in an instant. Among other factors, this hastened the end of the Cold War in America's favor. Thus, the original geopolitical goals of the Apollo program were once again realized, and in a manner undreamed of fifty years earlier.

The story becomes less definitive and not completely positive when evaluating Apollo's legacy to the idea of human spaceflight. During the era of the Apollo program, America learned how to journey in space with people and machines. The accumulation of such knowledge was not the result of any systematic attempt to acquire it for its own sake, but was developed and acquired because of need. The tight schedule dictated by a decadal deadline, coupled with the clear geopolitical need to demonstrate American technical superiority, made reaching certain technical milestones essential. We learned how to do orbital rendezvous because we needed to master that skill—and quickly. This lesson has been lost on many current space policymakers: the acquisition of true spaceflight capability results from the attempt to fulfill a mission, not from vague directives to "develop technology" so that we can eventually "go somewhere."

The Apollo program architecture—the legacy of launching a mission "all up" in one or two launches that deliver all the pieces needed for a single mission, discarding the expendable hardware along the way—persists in the minds of most space policy makers and planners to this day. While this approach worked for the fulfillment of Apollo's limited primary objective ("Man-Moon-Decade"), it is not conducive to developing a long-term, permanent spacefaring capability. The physics of spaceflight dictate that you use most of your rocket propellant to simply achieve low Earth orbit, with little, to no, fuel left to go beyond it. Apollo defied the "tyranny of the rocket equation"[17] through brute force, by launching a fully fueled Saturn IV-B stage that could throw some fifty-five tons along a translunar path. To go farther, or to go with more capability, requires either a much larger launch vehicle, multiple launches of a heavy lift vehicle, or the development of propellant depots in space. These depressing mathematics rapidly tally up to an infeasible launch rate, along with complex orbital operations needed to assemble an interplanetary craft. Yet, exactly such a cumbersome, impractical, and expensive approach is part of the current NASA Design Reference Mission for a human mission to Mars.[18]

For thirty years, following the end of Apollo, the enormous logistical requirements for sending human missions beyond low Earth orbit (LEO) made most manned space activity there unthinkable. In its place, other ideas began to emerge—concepts designed to take advantage of what space had to offer in terms of creating new capability from *what we could find out there*. Additionally, a more incremental approach was sought, whereby the pieces would be reusable, smaller, and less expensive. In part, the development of the space shuttle was pursued for these very reasons. Although the shuttle was not completely successful in obtaining this part of its various mission goals, the idea of an incremental program, developed using smaller, reusable pieces, remains attractive from a variety of perspectives.

The cost of the Apollo program still generates a lot of discussion.[19] The entire program cost an estimated $25 billion in 1965 dollars (about $200 billion in 2014 dollars). However, that number includes the construction from scratch of an enormous material infrastructure, such as the NASA field centers and the facilities used to test and stage the lunar missions. Much was made at the time about the "misplaced priorities" of the space program, as if the cancellation of Apollo would cure a plethora of social ills. Looked at from the perspective of ending the Cold War struggle with the Soviet Union, the race to the Moon was very cost-effective.

However, there was another aspect to Apollo, one that constrains our meaningful progress in space to this day. One of Apollo's baneful legacies was the entrenchment of the notion of exploration as a public spectacle or contest, designed to distract and excite the public. Although an attempt was made to justify the race to the Moon in terms of technical spinoff benefits, such efforts were always subject to the criticism that technical innovation would have occurred anyway, without a space program—an irrefutable proposition because the counterfactual cannot be demonstrated. Instead, supporters of ambitious space efforts have spent the last fifty years trying to convince policymakers that the country needs challenging and "exciting" goals to engage and inspire the public. This *panem et circenses* mindset remains a fixture of modern society and is an especially well developed standard used by the media to evaluate (and usually, denigrate) proposed new space initiatives.

This entrenched way of thinking is ineffectual and counterproductive. By making the human space program into an overblown "reality show," we are forever doomed to perform singular and unconnected stunts of no lasting value. Rationales for space exploration that require public "excitement" too often rely on being the "first" to do something. This involves promoting distant, unachievable goals such as human missions to Mars instead of reachable, near-term goals and destinations that we could accomplish on reasonable timescales, such as a lunar outpost. Current concepts of public support for space exploration are based on a false reading of public sentiment: Most people simply do not care about space, so attempts to "excite" them are bound to fail. There are always vocal proponents for space, individuals and small groups who hold strong opinions, but too often lack the necessary technical knowledge to understand what is feasible, against what they desire.

Despite this problem, a case still can be made that an affordable, long-term strategic goal for human spaceflight not only exists but can be adopted and attained without breaking the national bank. After fifty years of human spaceflight, we realize that there are tasks in space beyond the capabilities of robotic machines—tasks that require human presence. People must be present to interact physically and intellectually with the space environment in real time to accomplish some goals, such as scientific field exploration and the repair and maintenance of complex machines. We need to develop a system that ultimately permits us to go anywhere we need, with humans and machines, to accomplish whatever goals may be desired. A large ambition, to be sure, but we already have signs that the creation of such a space system is possible.

How? The answer is right next door.

3

After Apollo: A Return to the Moon?

The two decades following the end of the Apollo program are the wilderness years of lunar exploration. Despite repeated attempts and endless discussion, except for flybys by spacecraft on their way to somewhere else, between 1972 and 1994, there were no American missions to the Moon. Still, we continued to study the samples and data returned by the Apollo missions. There was occasional excitement when an international mission returned new lunar data or information, or when the odd American spacecraft acquired some new data as it flew past the Moon. During these years, we made significant advances in our understanding of the nature of the Moon and with it, gained a better understanding of the requirements for living there, all which added to the frustration of advocates desiring a return to the Moon.

The entire American space program suffered an identity crisis in the early 1970s. Following the success and hoopla of winning the race to the Moon, America seemed to lose interest in space. At least, that's what we were told had happened.[1] Social commentators decried the efforts of the American space program, describing them as irrelevant and a waste of money. Defenders of the space program spoke about technical spinoffs and societal inspiration. But the most inspirational aspect of Apollo turned out to be the most effective one used against it: the striking image of a nearly full

Earth rising above the lunar horizon, first seen during the Apollo 8 mission of December 1968. Similar images were subsequently captured by each succeeding mission. The view of a blue and white Earth suspended in black space above the barren, lifeless Moon initiated the modern environmental movement, which, in turn, quickly blossomed into a Luddite, anti-technological crusade. People were encouraged to eschew technology, forswear a modern civilized lifestyle, and go back to the land.

During this time, human spaceflight was focused exclusively on low Earth orbit. The development of the space shuttle was billed as a program that would "make spaceflight routine." Many equated "routine" with "cheap." While the program achieved the former, it did not attain the latter. With the space program effectively capped at less than 1 percent of the federal budget per year, there was no money to develop human missions beyond low Earth orbit (LEO). While the shuttle has been labeled a policy failure,[2] in truth, it offered several unique and valuable capabilities, including some that are not available now, or even contemplated to be present on any future manned spacecraft. The shuttle's development was marked by technical difficulties and fiscal challenges, but in hindsight, it is hard to see how it could have been done any better or more inexpensively.

In addition to developing the shuttle, NASA used surplus hardware from the Apollo Moon program to make Skylab, America's first orbiting space station.[3] Skylab was a Saturn third stage (S-IVB) with its interior configured into a living and laboratory space for three crewmembers to inhabit for periods of up to ninety days. The laboratory, launched on a Saturn V on May 14, 1973, quickly encountered problems when during ascent its thermal shield was torn away. Skylab also experienced significant problems when one solar array was torn off during launch, and the other did not deploy on arrival in orbit, pinned to the side of the lab and unable to generate electrical power. As a result, when the crew arrived a few days later, the workshop was severely underpowered and overheated. So severe were these problems that they threatened to cause an early termination of the first Skylab manned mission and the Skylab program as a whole.

Skylab 2's crew, consisting of Pete Conrad, Paul Weitz, and Joe Kerwin, went straight to work troubleshooting these problems. They erected a sunshade parasol that allowed the vehicle to remain cool under the glare of solar illumination. They conducted spacewalks to free the pinned solar array. Once it was fully deployed, it started producing electrical power. The crew spent a record-setting twenty-eight days in orbit, and thanks to their sustained and heroic efforts, Skylab was saved. During their long-duration mission, they activated on-board experiments, conducted a variety of medical experiments, mapped the Earth, and made solar observations with the use of a special telescope.

Two additional Skylab crews followed, spending periods of two and three months respectively in the orbiting space station. The last crew left the station in a configuration that would allow it to be visited and used by a future crew of the yet-to-fly space shuttle. In order to attach new solar arrays and a docking mechanism, and to outfit the laboratory for use by up to six or seven crew members, plans were developed to fly a couple of shuttle missions to Skylab in 1979–80. However, these missions never flew. The shuttle had run into development problems, which delayed its first launch until well after 1980. By the late 1970s, enhanced solar activity had heated and expanded the atmosphere outward, increasing the drag on Skylab. This increased drag made the orbit of Skylab decay at a much higher than anticipated rate, eventually leading to an uncontrolled reentry of the lab on July 11, 1979. Although NASA attempted to steer the vehicle to uninhabited ocean, large chunks of debris fell on the outback in southwestern Australia. Fortunately, no one was injured and there was little property damage.

Once believed to be the beginning of a long-term effort, one that would see Apollo space hardware conducting a wide variety of missions throughout cislunar space, Skylab now represented the shriveled remnant of our ambitious post-Apollo plans. By using the basic building blocks of Saturn and the Apollo command and lunar modules, this program, dubbed Apollo Applications, had envisioned space stations with orbital servicing vehicles, lunar orbital observatories, and even surface outposts. The problem for

Apollo Applications was that it needed the Apollo and Saturn production lines to remain open and that required more money than Congress and the President were willing to make available. The shuttle had been sold politically on the promise of making space affordable. Since there was no affordable way that both could be in production at the same time, something had to go. With the demise of the Apollo-Saturn production lines, plans for missions throughout cislunar space ended.

The space shuttle was originally designed to become the first piece of an entirely new line of reusable, extensible space hardware. The shuttle, as developed, could only go to and from low Earth orbit but its designers certainly had no intention of stopping there. The official name of the shuttle was the Space Transportation System (STS), a name chosen to convey that the Earth to LEO orbiter was only a single piece of a larger, more comprehensive system—a system that included a permanent space station and an orbital transfer vehicle, a space-based "tug" that could haul satellites and other payloads to high orbits above LEO. But that concept was gradually forgotten as we busied ourselves with specialized missions to LEO and with the monumental task of building a new space station—NASA's principal destination in space for the 1980s—assembled on orbit over time from pieces brought up by the shuttle.[4]

The end of the Apollo program was followed by the doleful coda of the Apollo-Soyuz Test Project (ASTP), a joint flight of American and Soviet human spacecraft designed to inaugurate a new age of cooperation in space and ensure peace on Earth.[5] The Apollo crew consisted of veteran Apollo astronaut Tom Stafford commanding, flying with Vance Brand and Deke Slayton, who was finally getting his chance in space after being grounded for thirteen years because of a heart murmur detected in 1962. They rendezvoused in space with a Soviet Soyuz spacecraft commanded by the first man to walk in space, Alexei Leonov, and his copilot, Valeri Kubasov. The two spacecraft docked using a common berthing mechanism provided by the United States. After exchanging handshakes and smiles, the crew drifted over the Earth in an extended demonstration of good will, good spirits, and fervent hopes for future cooperation in space. Cooperation would eventually come twenty years later after the Iron Curtain was dismantled, and with it the Soviet Union.

With the splashdown of the ASTP on July 24, 1975, America was without any means to send people into space until the new space shuttle system became operational. The shuttle was a complicated and delicate vehicle. It had to withstand a violent launch and ascent as well as a harrowing reentry speed of Mach 25, all while retaining a low enough mass to make the entire system work. Low-weight silica tiles that were glued onto the outside of the orbiter airframe provided the necessary thermal insulation to block the searing heat of reentry. These thermal protection tiles caused ongoing and endless headaches over the entire thirty years of shuttle operations. The thermal tiles were both fragile (likely to break if dropped) and tended to fall off the airframe (finding the right bonding agent to glue them in place took some time).

Several drop tests were conducted in which a shuttle was released from its carrier 747 aircraft and allowed to glide to the surface before the first space shuttle orbital mission launched in April 1981. Astronauts John Young and Bob Crippen flew the first shuttle orbiter *Columbia* into space and safely returned it to Earth.[6] That flight verified the system's basic design and started the next chapter in the history of the US space program. Shuttle flights continued apace throughout the 1980s, as flight after flight delivered satellites to their orbits and flew a variety of Earth observation and medical experiments. The shuttle design allowed it to be fitted with Spacelab, a cylinder-like module roughly the size of a school bus that was carried inside the shuttle cargo bay. Since Spacelab flew only during a shuttle mission, its operation in orbit was limited to about two weeks, the limit of the amount of reactants that could be carried for the shuttle's fuel cells and fuel for attitude control.

In addition to these high profile civil space missions, several early shuttle flights were dedicated to the launch of national security payloads. This was a consequence of promoting the shuttle to the Congress as the universal replacement for all expendable launch vehicles. The argument was made that shuttle could handle and deliver on orbit any and all payloads—scientific, commercial, and national security. Estimates made during vehicle development suggested that as many as fifty flights per year were possible. But after the shuttle became operational, the vehicle required much more refurbishment between flights (and consequently, more time to prepare for launch) than had been anticipated. At peak rates of activity, the shuttle flew about eight to nine times per year. While this flight rate was quite respectable for such a complex system, it was not the level of activity envisioned and desired in the early days of the program.

Despite the operational successes of the space shuttle program, a gradual sense of ennui developed within the space community. The program's seemingly endless series of missions to LEO had become its own justification, and it was perceived, perhaps unfairly, as a dead end. Initially, this was because there was no space station to support. When the additional pieces of the STS "system" did not materialize, it meant we had no station, no orbital maneuvering vehicle and no lunar tug. Thus, the STS had become a system with just one piece, and it was getting harder to justify a human space program that only orbited in endless circles.

The stasis, however, was more illusionary than real. Despite the shuttle program's focus on low Earth orbit, advanced program planners in Houston had indeed been thinking about follow-on steps once the orbiter became operational. In accordance with the classic von Braun architecture, the obvious next step was some type of space station.[7] Because Skylab had been lost and we no longer had the Saturn V launch vehicle, the shuttle would have to be used to assemble a space station. Using the shuttle as a delivery system meant that construction needed to be done in small pieces, with full station assembly requiring dozens of launches, spread out over many years.

Looking back from our vantage point of having an operational ISS, it is easy to forget what a monumental engineering and programmatic challenge that was. We had never assembled a giant, distributed-system satellite in space. The assembly of complex equipment and facilities in space would require techniques that had not yet been developed and were only vaguely understood. The assembly robots had not yet been conceived, let alone built, and a managerial structure had to be formulated that could adapt to changing budgets, module delivery schedules, weather delays, and pad availability. From their mid-1980s vantage point, those tasked with the challenge of assembling the station knew that many more unknowns than certainties lay ahead.

Having a large space station in low Earth orbit would offer more than just a laboratory for experiments: If properly constructed and configured, it could become the transit node for missions beyond low Earth orbit to the Moon and on to the planets. This idea still held sway in the minds of shuttle architects who took the moniker Space Transportation System literally. The von Braun architecture, laid out in the famous *Collier's* articles of the early 1950s,[8] envisioned first a space station, then an orbital transfer vehicle, a Moon tug and lander, and finally an interplanetary spacecraft. This incremental, building block approach had been abandoned when the political imperative of beating the Soviets to the Moon had taken center stage, but it was an approach to which the space agency wanted to return.

President Ronald Reagan announced plans for the new space station program in his 1984 State of the Union speech.[9] It would be called *Freedom* and would serve a variety of purposes, including laboratory research and observing the Earth and the universe, as well as serving as a transportation node. In its latter role, *Freedom* would be equipped with a servicing bay for satellite repair and serve as a departure point for missions from low Earth orbit to high orbits typically occupied by commercial communications and other satellites. Such transport required a reusable, refuelable vehicle, one that could move from low orbits to geosynchronous orbit (GEO), a circular, equatorial orbit about 22,000 miles (36,000 km) high. At this altitude, satellites orbit once every twenty-four hours and thus appear stationary or trace elongated, figure-eight loops in the sky. Any ground station on the hemisphere below the satellite in GEO is always in radio view. GEO is extremely important real estate for global communications, weather monitoring, and remote sensing.

A rocket launched from the surface of the Earth expends virtually all of its fuel to achieve low Earth orbit. But in terms of orbital energy, this is only about halfway to geosynchronous orbit. To get satellites to GEO, the rocket would have to carry an upper stage for the final orbital transfer, thus limiting the size, and therefore capacity, of a satellite in GEO. Additionally, a satellite in high GEO would not be accessible by the shuttle, or any other human spacecraft to date. When a high orbit satellite malfunctions, typically it is abandoned and deorbited, whereupon an entirely new satellite must be built and launched.

Having a spacecraft stationed at the low orbit space station would solve this dilemma. It would permit crews to travel routinely to and from the high orbits that these satellites occupy to service or replace them. But more significantly, crews could build satellite systems that would be much larger and more capable than any that could be launched on a single existing or planned launch vehicle. If the building of *Freedom* were successful, it would teach us how to build large, distributed systems in space. These techniques could then be applied to complex satellites in high orbits, presuming that there would be a way to get crews and repair facilities to that high orbit.

A key piece of the evolving STS, the projected orbital transfer vehicle (OTV), was designed to be berthed at *Freedom* and available to transport people and equipment to higher orbits when needed. The OTV was to be fueled by liquid hydrogen and oxygen brought up from Earth, at least initially. The vehicle also carried a heat shield, allowing it to use the friction of Earth's atmosphere to slow down the vehicle during close approach on return to LEO, thus requiring only a minimal amount of maneuvering capability. This strategy made the vehicle smaller and more efficient. Development of an OTV would be the next link in creating a genuinely space-based transportation system—and a vehicle that could routinely reach GEO could also go to the Moon.[10] Despite its many potential benefits, an OTV was never built.

The Lunar Base Movement (1983–93)

In 1983, two scientists from the Johnson Space Center, Michael Duke and Wendell Mendell, realized that if NASA developed the OTV as part of the shuttle-station architecture, we would possess the means to return to the Moon. Along with physicist Paul Keaton from Los Alamos National Laboratories, they organized a small workshop, which was followed by a major conference at the National Academy of Sciences in Washington.[11] That conference drew a large and enthusiastic attendance of engineers, scientists, and space visionaries. Over the course of three days, they discussed and pondered the implications of a lunar return. The scope of the meeting varied widely, with such topics discussed as extended exploration of the Moon, habitation and life support, mining and use of local materials for oxygen and construction, and orbit-to-surface transportation and fueling depots.

This meeting initiated a large community movement dedicated to lunar return. Enthusiasts and advocates studied and improved their knowledge of the lunar surface and materials in preparation of a return, not in the temporary, sortie mode of Apollo, but for longer, more permanent stays. A series of meetings, workshops, and conferences over the next few years fleshed out possible scenarios for lunar return. Much attention was paid to the possible use of lunar resources to support extended human presence on the Moon and elsewhere in space.[12] These schemes tended to focus primarily on the production of oxygen; lunar soil is about 45 percent by weight oxygen, although extracting it and converting it into its free, gaseous form was found to be a very energy intensive activity. Moreover, the environment of the low latitude regions of the Moon requires a long-lived source of electrical power in order to survive the fourteen-Earth-day-long lunar night. Thus, studies of power generation mechanisms needed at lunar outposts to keep equipment and people warm during the bone-chilling lunar night, revolved around the development of nuclear reactors, which would provide steady, constant electrical power and heat.

All these studies concluded that while lunar habitation was possible, it would require several expensive technical developments. Once again, space dreams ran up against the cold realities of fiscal constraints. The response to this realization tended to focus on justifying lunar return in terms of some high-value benefit, such that billions of dollars of investment would be worthwhile. Such benefits typically involved the production of clean electrical energy for the Earth. One idea was to make solar cells in situ on the lunar surface and create kilometer-sized photovoltaic arrays whose power output could be transmitted to Earth via microwave or laser. An alternative concept was to harvest the lunar regolith for a rare isotope of helium, ³He, which could fuel a "clean" fusion reaction, i.e., one that produces no harmful radioactive by-products.[13] Although ³He is present on Earth as a trace component of natural gas, it is found in extremely minute quantities, inadequate to fuel a commercial electrical generating industry. However, the Sun streams energetic particles continuously. This is the solar wind, which bypasses the Earth due to our global magnetic field but is implanted on lunar dust grains. Although still present in relatively minute amounts in the lunar soil (about twenty parts per billion), studies indicate that such concentrations are large enough such that ³He could be harvested from the Moon. This idea caught the imagination of both the public and the lunar return community when it was first proposed. However, several significant technical prerequisites remain before we have power generation systems that use ³He, most significantly, the need for a reactor design in which to burn the helium fuel.

A major activity of the lunar science community in the 1980s was an effort to send a robotic mission to orbit the Moon. This mission concept first emerged in the mid-1970s under the name Lunar Polar Orbiter (LPO), a perfect descriptor. Because the plane of an orbit is fixed in inertial space, a satellite in polar orbit will view the entire surface as the planet or moon slowly rotates on its axis. Such a spacecraft could be configured with nadir-pointing instruments to measure a variety of chemical, mineralogical, and physical properties of the Moon. All of the Apollo missions flew in near-equatorial orbits, so only about 20 percent of the lunar surface was overflown and mapped with compositional remote sensors from the orbiting Command-Service Modules. Both of the Lunar Orbiter IV and V spacecraft were placed in polar orbits and completed a global survey of the Moon, documenting their value.

One critical piece of information about the Moon was much debated in the years following Apollo. No evidence for water—past or present, on the surface or inside the Moon—was found in the lunar samples, a finding that led to the dogma that the Moon was bone-dry and that it had always been so. As such, this made the task of living on the Moon much more formidable and challenging. However, before the advent of the Space Age, we knew that the poles of the Moon had some unique properties. Because the spin axis of the Moon is nearly perpendicular (88.4°) to the plane of the ecliptic (the plane in which the Earth-Moon system orbits the Sun), the Sun always appears on or close to the horizon at the lunar poles. If you were on a peak, you could be bathed in constant sunlight. Conversely, if you were in a hole (crater) at the poles, you might never see the Sun (see figure 3.1). These dark areas would be extremely cold, since the only heat they receive comes from the extremely low quantities of heat flowing from the interior of the Moon itself.

Several studies suggested that these properties could have some dramatic consequences. We had evidence that the Moon has been bombarded by water-bearing objects—namely, comets and meteorites—over its history. Most of this water would be lost to space or dissociated in the high temperature vacuum of the lunar surface. However, if water somehow found its way into a dark "cold trap" near the poles, it would remain there forever,

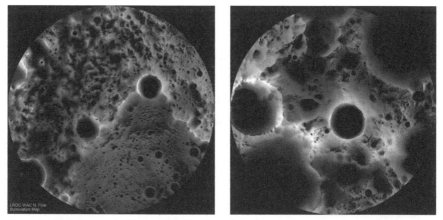

Figure 3.1. Lighting maps of the north (left) and south (right) poles of the Moon. On these composite images, bright areas are in sunlight for extended periods while black areas are in permanent darkness. This relation (caused by the 1.6° obliquity of the Moon) makes cold traps that have accumulated significant amounts of water ice over geological time. The sunlit areas permit electrical power to be generated nearly continuously.

and no known natural process could extract it. Much speculation was expended on how much ice might be in the polar regions of the Moon, but we could not know if it was there until we went looking for it.[14] A polar orbiting, remote sensing satellite (e.g., LPO) was needed to detect what might be in those dark areas.

Despite its appeal on scientific grounds, and its obvious importance as a precursor for eventual human return to the Moon, the LPO mission was repeatedly passed over for other missions throughout the twenty years following Apollo. In January 1986, the space shuttle *Challenger* exploded shortly after liftoff, killing all seven of its crew, including Christa McAuliffe, who was not an astronaut but instead the first teacher in space. The shock of this tragedy was a public relations disaster for the agency, followed by a wrenching period of introspection and soul-searching about its vision and purpose, along with the accompanying technical reviews called up to fix the problem and restart the shuttle program. In addition to agency chaos after the *Challenger* accident, the *Freedom* project was also in turmoil, having undergone two complete redesigns before the shuttle accident, followed by another redesign a year later. Because of the major disruption of the loss of a shuttle, serious concerns were raised about the viability of the space station program.

Two reports were issued during the manned spaceflight hiatus of the late 1980s. The Rogers Commission, named after its chairman, former Secretary of State William Rogers, was chartered to identify the cause or causes of the *Challenger* accident and to recommend policies and procedures to fix the problem.[15] The other commission had a broader task: The National Commission on Space (NCOS), also called the Paine commission after its chairman, former NASA Administrator Thomas Paine, was asked to devise a set of long-range goals for space and to identify some of the strategies needed to attain them.[16] The NCOS work was near completion when the *Challenger* accident occurred, and because of this unfortunate timing, its report was largely ignored when released. However, the Paine Commission report was very thorough and complete. It identified a systematic, incremental, and affordable expansion of humanity into space, for all the reasons we have identified over the years—the NCOS vision prominently featured space resource utilization in addition to exploration and science. It anticipated almost all of the current arguments for space goals and destinations, and suggested that because all are desirable in the objective sense and have their own constituencies, each can and should be pursued via a program that incrementally develops a wide range of capabilities.

The agency responded to the Paine report with a series of studies and workshops throughout the hiatus period in human spaceflight, culminating with a report issued in August 1987 by an internal study group led by astronaut Sally Ride. The Ride Report identified four mission concentration areas: Earth system science from space, unmanned space science exploration, a lunar outpost, and a human Mars mission.[17] The report did not advocate or choose any of the four but instead focused on what benefits and spacefaring legacies each one would give us. It suggested that a heavy lift launch vehicle would enable many of these activities and that a new HLV, using shuttle-derived hardware, could be developed quickly and inexpensively.

Rise and Fall of the Space Exploration Initiative (1989–93)

Problems with the shuttle solid rocket booster joints (identified by the Rogers Commission) were corrected and the vehicle returned to flight in September 1988. Armed with a renewed capability to get humans to and from orbit and with reports from three blue-ribbon study groups, President George H. W. Bush made the decision to announce a new major direction for America's space program. Much speculation has been expended on the origins of the subsequent Space Exploration Initiative (SEI).[18] My interpretation is simple: at the end of the 1980s and the beginning of the 1990s, as the Cold War was winding down in our favor, concern had developed about the erosion of our national technical capabilities—the enormous defense industrial infrastructure that won the struggle against the Soviet Union. President Bush and his advisors were well aware of this issue and the need to maintain a level of advanced technical infrastructure in the absence of the Cold War political imperative. The space program had served that purpose before and thus, an expanded space program—made affordable by the easing of defense requirements—could maintain a keen technological edge at a fraction of the level of Cold War defense expenditures. Curiously, the members of the Bush administration responsible for space policy never made this point publicly, but I know from discussions with some of them that many in the White House were well aware of its dimensions and implications.

In a special speech delivered on the steps of the National Air and Space Museum in Washington DC, President Bush announced the new initiative on the twentieth anniversary of the Apollo 11 Moon landing.[19] The SEI was what space enthusiasts had been wanting since the Apollo program: a presidential declaration on ambitious space goals. It called for the completion of space station *Freedom*, a return to the Moon ("this time to stay"), and a human mission to Mars. The president did not set forth deadlines for each milestone, except that space station *Freedom* should be completed within the next decade and that missions to the other destinations were tasks for the new millennium. The president asked his own White House National Space Council to examine and define the technologies and architectures needed to implement his new space initiative. Naturally, the Space Council turned to NASA for assistance in this new task.

Teams from NASA Headquarters and the field centers were quickly assembled and charged with defining the steps, and the missions and pieces of the new program. They were tasked to report to the White House within ninety days. The "90-Day Study" soon became infamous as the death certificate of the SEI, although in hindsight, it is not nearly as nefarious as widely reported and believed, and in fact, contains much good engineering sense and many clever ideas.[20] In short, the main problem was that NASA was barely being funded at an adequate level to run the space shuttle program and to build *Freedom*. Naturally, it would require additional funding if additional major tasks were added to its agenda. Such logic was forgotten or ignored in an orgy of self-righteous indignation over the "pedestrian and bloated approach" of the 90-Day Study. Five alternative "reference approaches" were outlined, with each building outward from the shuttle/ station in incremental steps while varying the rate of development and the amount of activity according to selectable levels of effort.

The biggest problem with the 90-Day Study was not the report itself, but what happened behind the scenes. The report deliberately did not include budget information. Estimated costs were prepared so that policymakers could evaluate differences among the approaches. As one might expect, once these cost numbers were leaked to the press, the chattering classes inside the Beltway were aghast: the new SEI was expected to cost upward of $500 billion! What was always left out of these stories was that this cost number was the aggregate budget of the agency spread over the course of thirty years, a metric against which few federal agencies would stand up well under scrutiny. And given the national security dimension of the new SEI, such sums were a mere fraction of the national defense budget over the same period. Nonetheless, this number was widely circulated. It quickly became "canonical" and was used to discredit and disparage the whole idea of the SEI.

The White House and Space Council knew that political forces outside their control were torpedoing their new, major initiative.[21] To fight this effect, the Space Council convened a special committee to examine the 90-Day Study, as well as to review detailed alternatives prepared by industry and other federal entities. The most famous of the latter was the proposal from Lawrence Livermore National Laboratory to use inflatable vehicles launched on existing expendable rockets.[22] This proposal claimed that both a lunar return and a manned Mars mission could be conducted for less than one-tenth the leaked cost of the 90-Day Study. Regardless of the doubtful veracity of that cost estimate, or the technical feasibility of the concept, it drew major attention from the White House. That attention propelled the canvassing of a wider segment of the community with hopes it would generate new and innovative ideas with which to implement the SEI for a fraction of the funding that NASA claimed was needed. The National Research Council, whose special report on the study concluded that a variety of other technical options should be investigated, ones that NASA had not considered, provided additional support for a major reevaluation of the 90-Day Study.

In this vein, the Space Council decided to create an outreach effort that would gather up the best technical ideas on how to implement the SEI from all sectors. These educated and innovative suggestions and plans were to be collected, evaluated, and high-graded by a special panel called the Synthesis Group and distilled into a plan for a magical—meaning cheap—beanstalk into space. This panel included members from academia, government, and industry and was chaired by astronaut Tom Stafford. I was a member of this group from August 1990 to June 1991. Tom Stafford said this activity was "like drinking from a fire hose," and I found that to be an apt description. The ten months spent serving on Synthesis was a crash course in astronautics, a course that included the benefits and pitfalls of technology development and its role in architectural design. As one might expect, the massive input from the space community did not contain any "magic beans" or "silver bullets" that would take us to the Moon and the planets faster, better, or cheaper.[23] And in that sense, the Synthesis Group did not succeed. But in another sense, the Synthesis Group advanced our understanding about the Moon and its crucial role for human expansion into the solar system.

Two events occurred in the spring and summer of 1990 that severely damaged the cause of SEI. The first event involved the Hubble Space Telescope. Although the telescope had been successfully launched, it was soon discovered that its main optical element had been ground to the wrong specification. This mistake caused Hubble's highly anticipated new images of the universe to be out of focus.[24] The other event was the temporary grounding of the shuttle fleet because of an unresolved hydrogen leak. These problems, along with the release of the 90-Day Study, combined to present the image of a space agency that was both technically incompetent and politically out of touch. Thus, despite giving the space agency an additional $2 billion overall in the FY 1991 budget, Congress zeroed out the SEI, a clear signal that NASA was in serious political trouble. The deep antipathy between the space agency and the White House was finally resolved with the sacking of Richard Truly as administrator and the subsequent hiring of Daniel Goldin as his replacement. Despite attempts to initiate SEI again in the following two years, Congress would not approve or fund it, and the initiative was terminated following the reelection defeat of President Bush and the advent of the Clinton administration.[25]

The Clementine Mission and Its Legacy (1994)

The lunar science community continued to lobby NASA to send a robotic orbiter to the Moon, but to no avail. Their goal was to map the Moon's shape, composition, and other physical properties. Such a mission would not only document the processes and history of the Moon but would also serve as an operational template for the exploration of other airless planetary objects. A collection of global remote sensing data could provide scientists with invaluable ground truth when used in conjunction with the previously returned Apollo surface samples. The Lunar Polar Orbiter mission, proposed several times, never received a new start. Its last incarnation was the Jet Propulsion Laboratory's Lunar Observer, patterned after the ill-fated Mars Observer mission. The cost review of Lunar Observer came in at around $1 billion in 1990 dollars. Of course, it was passed over yet again.

Stewart "Stu" Nozette of Lawrence Livermore National Laboratory, another Synthesis Group member, was involved in the Brilliant Pebbles (BP) program of the Defense Department's Strategic Defense Initiative.[26] The idea behind BP was to defend the nation against ballistic missiles by launching swarms of small, inexpensive satellites, each capable of observing, calculating and plotting an intercept course to incoming missiles (the "brilliant") and then rendering them inoperative by collision (the "pebble"). These small, three-axis stabilized vehicles carried imaging sensors (both active and passive) as well as in-flight computers and propulsion systems. In short, they were small but fully capable, self-contained spacecraft.

Nozette's idea was to fly a BP to a distant target in space. Because of his interest in space resources, he devised a mission that would fly by an asteroid and possibly orbit the Moon. Stu and I discussed these possibilities, and it seemed that a fairly significant mission might be built around these small spacecraft. My colleague Eugene Shoemaker of the US Geological Survey was brought in early on the planning of this mission. Gene was a legend in planetary science circles. A member of the National Academy of Sciences, he had done the original geological mapping of the Moon before the Apollo program and was actively researching asteroids. His interest and involvement with the mission brought both prestige and credibility to the idea.

An agreement between NASA and the Strategic Defense Initiative Organization (SDIO) specified that NASA would provide the science team and the communications tracking support for the flight, and that SDIO would provide the sensors, spacecraft, and launch. The sensors had been developed at Livermore as part of the BP program, while the Naval Research Laboratory (NRL) would design and build the spacecraft, later named Clementine. Launch was on a surplus Air Force Titan II rocket, the same vehicle NASA used to launch the two-man Gemini missions in the 1960s. Because the Titan II pad at the Cape had been dismantled, the mission would be launched from Vandenberg Air Force Base near Lompoc, California.

The mission would put Clementine in a polar orbit around the Moon for two months, providing global coverage. The spacecraft would map the color of the lunar surface in eleven wavelengths in the ultraviolet, visible, and near-infrared portions of the spectrum and measure the Moon's shape from laser ranging. Other remote measurements would be acquired as opportunity presented. After this phase, Clementine was to leave lunar orbit and fly by the near-Earth asteroid Geographos. Program Manager Pedro Rustan, an Air Force colonel, was a skilled, tough engineer who kept us to deadlines. Stu became his deputy, coordinating many different activities, ranging from science objectives to spacecraft fabrication and testing. The Science Team, twelve lunar scientists with varied expertise, was selected from individual proposals submitted to NASA. Gene Shoemaker was named the team leader, and I was his deputy. Together, we planned mission operations with the NRL and Livermore teams. The Science Team carefully selected the filter bandpasses for the imaging systems that would allow the identification of lunar rock types from the color images.

The Clementine mission was remarkable for its short development cycle and cost. Twenty-two months elapsed from project start to launch, while a typical NASA planetary mission took from three to four years. In FY 1992 dollars, NRL spent about $60 million for the spacecraft and the mission control center. Livermore spent about $40 million on support services and on the production of the mission sensors. The Titan II launch vehicle and services, supplied by the Air Force, were valued at about $20 million, with an additional $10 million or so for avionics upgrades. The NASA Science Team cost a couple of million dollars, and the Deep Space Network support was a few million more. By totaling those numbers, I estimate that the mission cost about $140 million, or $540 million in today's dollars; for comparison, the then-recently lost NASA JPL Mars Observer mission cost a bit over $800 million, or more than $2 billion in 2014 dollars.

Those cost numbers caused considerable controversy, with some in the scientific community whining that the massive "Star Wars" (SDI) program absorbed and hid much of Clementine's cost. In fact, the whole point of the Brilliant Pebbles program was to adapt cheap, rugged tactical sensors to deep space use and thus take advantage of the cost savings provided by mass production (as opposed to the custom builds of most space systems). Moreover, there was nothing to stop NASA from using this same technology, other than a not-invented-here mindset and the still-prevalent tendency in the space science community to gold-plate scientific payloads.

The Clementine mission demonstrated the value of the so-called Faster–Better–Cheaper (FBC) paradigm.[27] The concept is not that cheap missions are inherently "better" but that by carefully restricting mission objectives to only the most essential information, it is possible to fly smaller capable missions that can return 80 to 90 percent of the most critical data; resources are often squandered in an attempt to achieve that last 10 percent of performance. Maybe FBC should be renamed Faster–Cheaper–Good Enough. The broad success of NASA's Discovery program over the last twenty years, in which mission objectives are carefully defined and limited to control overall cost, is testament enough to the general validity of the FBC concept. In addition to its scientific return, the Clementine mission flight-tested and qualified twenty-two new spacecraft technologies, including solid-state data recorders, nickel-hydrogen batteries, lightweight components, and low-mass, low-shock, nonexplosive release devices. All of these technologies have been employed on dozens of subsequent space missions, making many of these spacecraft lighter, more reliable, and longer-lived.

On the morning of January 25, 1994, less than two years after project start, the members of Clementine's science team stood together on a cold, windy California beach just a couple of miles from SLC-4W. We watched as the Clementine Titan II rose above the launch pad on a cloud of orange smoke and flame, arching into the clear, blue Pacific sky. We followed the vehicle's progress all the way through staging before losing sight of it. I left Vandenberg excited about the mission ahead, but my mood quickly changed when Science Operations Manager Trevor Sorensen sent news that we were in danger of losing the spacecraft (erroneous commands had been sent to Clementine, and the spacecraft was out of control). Fortunately, we recovered.

Once our spacecraft had safely inserted itself into orbit around the Moon and began mapping its surface, we were eager to get our first images. Our perch for receiving this mission data was a converted National Guard armory in Alexandria, Virginia. Dubbed the Batcave, the armory served as mission control center for the duration of the mission. Designed to save fuel, Clementine had taken a month-long, leisurely looping trip to the Moon, arriving there on February 19. When the first image finally flashed on the screen, I immediately recognized the crater but due to all the excitement, initially drew a blank on its name. Quickly consulting the wall map, I saw that we were looking at Nansen, a crater located near the north pole. A very strong sense of physically being present at the Moon came over me—I was flying across a landscape as familiar to me as any one that I knew on the Earth.

Mission operations became a regular series of work cycles arranged around the routine of collecting and downlinking data, verifying that the data was good, and making some initial scientific observations, although a couple of incidents from my time in the Batcave stand out.

As Clementine's orbit was about to pass over Tycho, the largest rayed crater on the near side of the Moon, I alerted everyone in the Batcave's control room that something incredible was about to appear. Audible gasps greeted the spectacular images of the floor and central peak of Tycho that came into view. On another occasion, Dave Smith, a science team member from NASA–Goddard Space Flight Center, asked how much polar flattening might be expected for the Moon. I replied "almost none," mainly because of the slow rotation rate of the Moon (once every 708 hours) combined with the rigid, nonplastic state of the lunar globe. Then, as the orbital ground tracks slowly marched westward across the far side of the Moon, we saw an astonishing falling off of topography toward the south pole. This large negative relief was the rim and floor of the South Pole–Aitken (SPA) basin, an impact crater more than 2,600 kilometers across and more than 12 kilometers deep. Geologists had long known that this basin was present, but until Clementine mapped its topography, no one had fully appreciated its huge size and state of preservation.

By now, Clementine had already shown us the nature of the polar regions of the Moon, including peaks of near permanent sun-illumination and crater interiors in permanent darkness. From his first look at the poles, Gene Shoemaker had an inkling that something interesting was going on. Gene tried to convince me that water ice might be present there, an idea about which I had always been skeptical. At that time, no trace of hydration had ever been found in lunar minerals, and the prevailing wisdom was that the Moon had always been bone-dry. With Gene arguing for us to keep an open mind and Deputy Program Manager Stu Nozette devising a bistatic radio frequency (RF) experiment to use the spacecraft transmitter to "peek" into the dark areas of the poles, we moved ahead on planning our observations. This turned out to be the setup for a history-making event midway through the orbital mapping campaign.

Although Clementine did not carry sensors for the detection of water, Stu believed we could improvise an experiment using the spacecraft's radio transmitter to "look into" the dark (and thus very cold) areas near the poles, places where water ice might exist. Radio echoes from the Moon could be detected on the giant radio antenna dish at Goldstone in California's Mojave Desert. With careful planning and commanding of the spacecraft by Radio Engineer Chris Lichtenberg, we successfully took bistatic radio frequency (RF) data of both poles during those phasing orbits, when Clementine shifted the perilune (low point) of its polar orbit from 30° south to 30° north latitude.

To my astonishment, a single pass over the dark areas of the south pole of the Moon showed evidence for enhanced circular polarization ratio (CPR), a possible indicator of the presence of ice. A control orbit over a nearby sunlit area showed no such evidence. However, CPR is not a unique determinant for ice, as rocky, rough surfaces and ice deposits both show high CPR. It took a couple of years for us to reduce and fully understand the data, but the bistatic experiment was successful—and a huge scientific bonus. In part, our ice interpretation was supported by the then-recent discovery of water ice at the poles of Mercury (a planet very similar to the Moon with a comparable polar environment).[28] Our published results in a December 1996 issue of *Science* magazine set off a media frenzy, followed by a decade of scientific argument and counterargument about the interpretation of radar data for the lunar poles—an argument that continues to a lesser degree to this day, despite subsequent confirmation of lunar polar water from several other detection techniques.[29]

The Batcave played host to several distinguished visitors during the two months that Clementine orbited the Moon, most notably astronauts John Young, a familiar face to members of the Synthesis Group and always a friend of lunar science, and Wubbo Ockels, a Dutch physicist with the European Space Agency. Ockels, encouraged by Clementine's success, campaigned to generate enthusiasm for small, cheap lunar missions at ESTEC, the European space center in the Netherlands where he worked. US Representatives Bob Zimmer and Jim Moran were impressed with our operation and pledged their support in Congress for future space efforts like Clementine. Finally, then-new Administrator of NASA Dan Goldin visited, distributing lapel pins and offering encouragement to the worker bees. Not all at NASA were enamored with the mission though, with some resenting the attention it had drawn, particularly with regard to the inevitable comparisons with their own ongoing and budget overrunning robotic missions.

With Clementine, we had successfully returned to the Moon, mapped it globally, and made several significant discoveries. A Science Team press conference was scheduled at NASA Headquarters to report on the new scientific findings, but NASA intervened at the last minute and cancelled our briefing. Several mutually exclusive excuses were given for this cancellation, but it was clear to members of the science team that some in the agency wanted to keep a lid on the scientific success of the mission, which was embarrassing to NASA because Clementine was much cheaper than similar agency efforts, yet just as scientifically productive, if not more so. But in time, news of the discovery of "the most valuable piece of real estate in the solar system" was revealed. With urging from the planetary science community, NASA agreed to fund a research program to take advantage of the abundant new lunar data acquired by Clementine.

Two cameras on Clementine with eleven filters covered the spectral range of 415 to 1900 nm, where absorption bands of the major lunar rock-forming minerals (plagioclase, pyroxene and olivine) are found. Varying proportions of these minerals make up the suite of lunar rocks. Global color maps made from these spectral images show the distribution of rock types on the Moon. The uppermost lunar crust is a mixed zone, whose composition varies widely with location. Below this zone is a layer of nearly pure anorthosite, a rock type made up solely of plagioclase feldspar—the original lunar crust, formed during the global "magma ocean" melting event. Craters and large basins act as natural "drill holes" in the crust, exposing deeper levels of the Moon. The deepest parts of the interior (and possibly the upper mantle) are exposed at the surface within the floor of the enormous (2,600 km diameter) South Pole–Aitken basin on the far side of the Moon.

Before Clementine, good topographic maps existed only for the near-equatorial areas under ground tracks of the orbital Apollo spacecraft. From Clementine's laser ranging data, we obtained our first global topographic map of the Moon. It revealed the vast extent and superb preservation state of the SPA basin and confirmed many large-scale features, mapped or inferred, from only a few clues provided by isolated landforms. Correlated with gravity information derived from radio tracking, we produced a map of crustal thickness, thereby showing that the lunar crust thins out under the floors of the largest impact basins.

As a result of this mapping, scientists could place the results of studies of the Apollo samples into a regional, and ultimately, a global context. Clementine collected special data products, including broadband thermal, high resolution and star tracker images for a variety of special studies. In 1996, after our paper was published in *Science*, a press conference was held at the Pentagon to announce the results of the bistatic experiment: the discovery of ice at the south pole of the Moon. In addition to discovering new knowledge of lunar processes and history, this mission led a strong wave of renewed interest in the processes and history of the Moon, an interest that spurred a commitment to return there with both machines and people. By peeking into the Moon's dark polar areas, we now stood on the edge of a revolution in lunar science.

This renewed interest in the Moon led to the selection of Lunar Prospector (LP) as the first of NASA's new, low-cost Discovery series of planetary probes. This mission found enhanced concentrations of hydrogen at both poles, again suggesting that water ice was probably present there. Buttressed by this new information, the Moon once again became an attractive destination for robotic and human missions. With direct evidence for significant amounts of hydrogen (regardless of form) on the surface, there now was a known resource that would support long-term human presence. Lunar Prospector's hydrogen discovery was complemented by the identification in Clementine images of several areas near the pole that remain sunlit for substantial fractions of the year—not quite the "peaks of eternal light" anticipated by the astronomers Beer and Mädler in 1837, but something very close to it.[30] The availability of material and energy resources, the two most pressing necessities for permanent human presence on the Moon, was confirmed in one pass. These two missions certified the possibility of using lunar resources to provision ourselves in space, thus permanently establishing the Moon as an enabling asset for continued human spaceflight. A remaining task was to verify and extend the radar results from Clementine and to map the ice deposits of the poles.

Missions flown over the last twenty years show how significantly Clementine's programmatic template has influenced spaceflight. The Europeans flew the SMART-1 spacecraft to the Moon in 2002, largely as a technology demonstration mission with goals very similar to those of Clementine. NASA directed the Applied Physics Laboratory (APL) to fly the Near-Earth Asteroid Rendezvous (NEAR) spacecraft to the asteroid Eros in 1995 as a Discovery mission, to attain the asteroid exploration opportunity missed when control of the Clementine spacecraft was lost after leaving the Moon; the mission was renamed NEAR-Shoemaker after Gene's tragic death in an automobile accident in Australia in 1997. India's Chandrayaan-1 had a size and payload scope similar to Clementine. The selection of the LCROSS impactor as a low-cost, fast-tracked, limited objectives mission further extended use of the Clementine paradigm.

The Faster-Better-Cheaper mission model, once panned by some in the spaceflight community, is now recognized as a valid mode of operations, absent the emotional baggage of that name.[31] A limited-objectives mission that flies is more desirable than a gold-plated one that sits forever on the drawing board. While some missions do require significant levels of fiscal and technical resources to attain their objectives, an important lesson of Clementine is that for most scientific and exploration goals, "better" is the enemy of "good enough." Space missions require smart, lean management; they should not be charge codes for feeding the beast of organizational overhead. Clementine was lean and fast; perhaps we would have made fewer mistakes had the pace been a bit slower, but despite its shortcomings, the mission gave us a large, high-quality dataset, one still used extensively to this day. In recognition of its substantial accomplishments, the Naval Research Laboratory transferred the Clementine engineering model to the Smithsonian in 2002, where it was put on display in the National Air and Space Museum, suspended above the display of the Apollo Lunar Module.

It is probably not too much of an exaggeration to say that Clementine changed the direction of the American space program. After the failure of SEI in 1990–92, NASA was left with no long-term strategic direction. For the first time in its history (but alas, not the last), the agency had no follow-on program to the shuttle/station, despite attempts by Dan Goldin and others to secure approval for a human mission to Mars, an insurmountable challenge both technically and financially. This programmatic stasis continued until 2003, when the tragic loss of the space shuttle *Columbia* led to a top-down review of US space goals. Because Clementine had documented the strategic value of the Moon, the lunar surface once again became an attractive destination for future robotic and human missions. The resulting Vision for Space Exploration (VSE) in 2004 made the Moon the centerpiece of a new American effort beyond low Earth orbit. Though Mars was declared as an eventual (not ultimate) space objective, specific activities to be done on the Moon were detailed in the VSE, particularly with regard to the use of its material and energy resources to build a sustainable program. Regrettably, as I will detail, various factors combined to subvert the Vision, thereby ending any strategic direction for America's civil space program.

Clementine was a watershed, a hinge point that forever changed the nature of space policy debates. We now recognize a fundamentally different way forward in space—one of extensibility, sustainability, and permanence. Once an outlandish idea found in science fiction, we now know that lunar resources can be used to create new capabilities in space, a welcome genie that cannot be put back in the bottle. Americans need to ask why their national space program was diverted from such a sustainable path. We cannot afford to remain behind while others plan and fly missions to understand and exploit the Moon's resources. Our path forward into the universe is clear. In order to remain a world leader in space, and a participant in and beneficiary of a new cislunar economy, the United States must again direct its sights and energies toward the Moon.

4

Another Run at the Moon

With the completion of the successful Department of Defense Clementine mission, the Moon was again viewed as a destination of value. Both Lawrence Livermore and the Naval Research Laboratory prepared for a Clementine II asteroid flyby mission, the unmet objective of the first mission due to a technical failure with Clementine's thrusters after it departed the Moon. Some on the study team proposed a mission profile that would mirror the plan of the first Clementine, an asteroid flyby followed by insertion into lunar orbit. The objective would be to map the Moon at greater resolution with additional instruments and follow up on the discoveries made by the first Clementine. This renewed lunar attention was not without detractors, who questioned what Clementine had found. Congress appropriated funds for the mission in 1997, but plans to fly it were scuttled when President Bill Clinton used his newly acquired line-item veto to zero out funds for Clementine II. The Supreme Court subsequently declared the line-item veto unconstitutional, but too late to save the Clementine II mission.

Around this time, NASA called for proposals to the Discovery program, a new series of small planetary missions, cost-capped at $150 million.[1] This mission series was NASA's attempt to emulate the Faster–Cheaper–Better paradigm that Clementine encapsulated. The new NASA administrator

Dan Goldin was renowned for his advocacy of the FBC mode of business. The Discovery program received dozens of mission proposals. A single planetary scientist, called the principal investigator (PI), led each proposal. In 1995, NASA picked Lunar Prospector (LP), led by Alan Binder, as the first Discovery mission, deeming it the least expensive, least risky mission proposal it had received, requiring only a small operations team. Its selection also avoided having the Clementine team again set foot on what NASA thought to be its turf, since a second NRL Clementine multiple asteroid flyby of comparable cost was also proposed.

The particle and geochemical sensors of LP perfectly complemented the multispectral images and laser altimetry data obtained by Clementine. Combined, these two missions gave us our first global look at lunar mineral and chemical compositions, surface topography and gravity, and regional geology and produced the data sets of the never-flown Lunar Polar Orbiter, the mission scientists had long desired. For example, we found that high concentrations of radioactive elements in the lunar crust are localized in the Procellarum topographic depression of the western near side, an unusual global asymmetry that is still unexplained. More importantly, LP's neutron spectrometer found high concentrations of hydrogen at both poles in roughly equal quantities. The neutron experiment only measures the concentration of elemental hydrogen, not its physical state—that is, whether it is present in the form of water ice or excess solar wind gas in the cool polar regolith. From this information, as well the results on lunar polar lighting and the bistatic radar from Clementine, the evidence continued to mount that something very interesting was present at both poles of the Moon.[2]

The Moon's spin axis is inclined 1.5° from the normal to the ecliptic plane, a nearly perpendicular orientation; this means that the Sun always hovers near the horizon at the poles of the Moon. The apparent angular width of the Sun at the Earth-Moon distance is about 0.5° of arc, so sometimes the Sun could be above the horizon and at other times, below it. But because the Moon's surface is rough and irregular, with large craters and basins, there are areas near the poles that in theory could see either permanent sunlight or permanent darkness. Clementine spent only seventy-one days in lunar orbit during the southern winter solstice, so it provided illumination data for only part of the lunar year. Nonetheless, analysis showed that small areas near the crater Shackleton, located near the south pole of the Moon, were sunlit more than 70 percent of the southern winter day. Three areas near the north pole were illuminated 100 percent of the day (northern summer). Data for the opposite seasons were not obtained.

An intense scientific debate over the existence of ice at the lunar poles spanned most of the decade around the turn of the millennium. The lunar ice controversy stemmed from the ambiguity of radar CPR as an indicator of both surface physical properties and composition. Because the data were nondeterminative, they proved fertile ground for intense argument. A paper presented at the 1995 Lunar Science Conference in Houston described the results of high resolution imaging of the lunar south pole by the large dish antenna at Arecibo, Puerto Rico. These images were able to peek into the totally sunless areas near the pole. Interestingly, small regions of high diffuse backscatter were seen.[3] This high diffuse backscatter, called circular polarization ratio, or CPR, is consistent with a surface composed of water ice, a high concentration of angular blocks, or both. Based on our result from the bistatic experiment, the Clementine team preferred the water ice interpretation, while others in the radar planetary science community argued for an origin from surface roughness.

The new LP neutron data clearly showed an excess of hydrogen at the poles, but the surface resolution of its hydrogen concentration maps was very low and we could not be certain whether the signal was caused by a large area of relatively low concentration—that is, solar wind gases implanted in the regolith—or by small, isolated zones of very high concentration such as ice in permanently dark areas. This controversy raged on as we tried to design, build, and fly a small imaging radar to the Moon to follow up on the Clementine and LP discoveries. Despite proposals for small NASA robotic missions, European Space Agency interest, and even some proposed commercial missions, no flight opportunities were to arise until late in 2003.

The greatest remaining unknowns were about the poles, those areas where we had found permanent darkness, possible permanent sunlight, and enhancement of hydrogen concentration, possibly indicating the presence of water ice in the dark regions. All of these new insights showed that the Moon was more complex and interesting than we had thought. A key discovery was the zones of extended sunlight. Finding areas on the surface that receive illumination for almost all of the lunar day removed one of the biggest hurdles to human habitation of the Moon: the need to provide a power source for electricity and heat during the fourteen-day nighttime. Nuclear power is best suited to the task, but the high costs of such power, both technical and societal, made lunar return unaffordable.

In contrast, the discovery of areas where power could be generated constantly by solar arrays now made extended stays on the Moon by people feasible. In addition, illuminated terrain—even by sunlight at grazing incidence—makes the extremely cold lunar night tolerable. Areas of constant sunlight near deposits of water ice create "oases" near the poles of the Moon where human habitation is possible and perhaps even profitable.

There was an interesting coda to the Lunar Prospector mission. After lowering its orbit to about 20 kilometers, as close as it is possible to orbit the Moon without running into some of its higher mountaintops, and collecting some high-resolution data from this close orbit, the spacecraft was deliberately crashed into a crater near the south pole on July 31, 1999. The objective of this effort was to kick up material from the impact to try to detect the polar water in the ejecta cloud with telescopes on Earth. Unfortunately, no ejecta were observed, so the debate over lunar polar water continued. (The same experiment was repeated ten years later during the LCROSS mission, with more productive results.) The LP spacecraft did carry an unusual cargo, however: some of Gene Shoemaker's ashes.[4] The urn and the LP spacecraft now rest in the floor of a crater near the south pole, subsequently given the name Shoemaker, a fitting tribute to Gene and his contributions to the study and exploration of the Moon.

We believed that we had found water at the poles of the Moon. Now we needed a commitment to go back to verify the new findings.

Mars Mania

Robotic missions to Mars have dominated the last twenty years of planetary exploration, an emphasis stemming in part from the planetary science community's efforts to fly a series of robotic missions that will eventually lead to the return of samples from Mars to Earth, an ambitious and very expensive proposition. With NASA's robotic spaceflight program still under intense scrutiny after the failures of the initial Hubble Space Telescope mission and JPL's 1993 Mars Observer spacecraft, Administrator Dan Goldin, a booster of both human Mars missions and the search for life, was unable to convince the Clinton administration or Congress to pony up enough money to fund an ambitious Mars sample return effort.[5] Undeterred, Goldin applied the FBC paradigm to Mars missions and moved forward with the Mars Pathfinder mission, a small rover called Sojourner that landed on Mars using a parachute and airbags deployed for final impact. Sojourner took some images and made a rudimentary chemical analysis of the soil. Pathfinder, although technically successful, did not fundamentally advance our knowledge and understanding of Mars and its surface processes and history, thus missing out on the "better"—or even the "good enough"—part of FBC.

An event that would concentrate everyone's attention on Mars during the 1990s turned out not to be from any space mission but from a laboratory right here on Earth. We had known for some time that a rare group of meteorites, the Shergottite-Nahklaite-Chassignite group (SNC), had unusual chemical properties and relatively young ages of crystallization. Analysis of these rocks found trapped argon within them identical in composition to analysis of the martian atmosphere made in 1976 by the Viking spacecraft. Scientists concluded that these meteorites are pieces of the martian crust, thrown into space by an impact on Mars that eventually made their way to Earth. The composition of these meteorites seemed to fit what we had inferred from Viking and remote sensing about the composition of Mars. These meteorites had ages much younger than most meteorites, nearly all of which formed 4.6 billion years ago, and lunar samples three to four billion years old, again congruent with our understanding of the extended geological evolution of the martian surface; crater density data suggested that ages of geological units on Mars spanned an estimated range from four billion to less than one billion years old. One rather unusual martian meteorite, ALH 84001, was found in Antarctica and determined to be relatively old: 4.5 billion years. Using a scanning electron microscope, tiny lifelike shapes were found inside the martian meteorite, forms resembling certain types of terrestrial bacteria, though much smaller and unique in detail. The authors of the ALH 84001 study suggested that these objects might be fossils of bacteria from an early epoch of Mars history. In other words, they asserted that traces of former extraterrestrial life had been discovered, a sensational claim that grabbed and held headlines.[6]

A flood of media coverage followed, eventually leading to press conferences at NASA Headquarters and finally, a Presidential Rose Garden statement. With that explosion of publicity, Dan Goldin moved to leverage some high-level political backing for a permanent, sustained Mars exploration program.[7] Although human missions to Mars remained beyond the reach of technology, a series of robotic probes leading up to that elusive sample return mission would keep the Mars scientists and NASA's Jet Propulsion Laboratory busy. "The Quest for Life" scientific gravy train was born.

By Saganizing the nation's civil space program—that is, by enshrining the Quest for Life as NASA's principal rationale for space exploration— Dan Goldin took the martian aspect of this new rationale and encapsulated it into the slogan, "Follow the Water." The idea behind this messaging was that life as we know it requires the presence of liquid water.[8] This dictum was followed by conducting a series of missions to areas on Mars where it was suspected that flowing water had occurred in the past. A cynical observer might notice that aside from the potential finding of extant or fossil martian life, no criteria for a programmatic exit from this exploratory path were defined. In essence, the new series of Mars missions took on a life of its own, becoming a permanent scientific and engineering entitlement program—a constant, uninterruptible cadence, elevated beyond the realm of peer-review selection pressure or second thoughts from planetary scientists, a situation viewed with dismay by the lunar community.

Despite the positive indications acquired from the Clementine and LP missions and subsequent studies of the presence of water ice and near-permanent sunlight near the poles of the Moon, NASA had no interest in conducting any follow-up investigations. Although these discoveries would permit long duration stays on the Moon and possible resources for additional spaceflight options, the lunar community could not get the funding to even *study* new missions that would have cost a small fraction of the JPL Mars program budget. Clearly, a policy decision had been made at some level that no large-scale human exploration program beyond LEO was possible: NASA's robotic exploration budget was to be sequestered over the next couple of decades for various Mars missions. This policy was never formally written down, but as they say, money talks. Predominant in every submitted NASA budget for robotic science during the Clinton years were "green" spacecraft for Missions to Planet Earth and Mars orbiters and rovers. The former were the pet project of Vice President Al Gore, while the latter was the implementation of the wish lists of Dan Goldin, Carl Sagan, the Jet Propulsion Laboratory, and likeminded members of the Sagan-founded Planetary Society.

However, the Mars exploration program started running into some serious technical problems. After the renowned Pathfinder mission in 1997, the first successful mission to Mars since Viking two decades before, the next two Mars missions failed. The Mars Climate Orbiter, a 1999 mission designed to characterize atmospheric phenomena and look for possible clues to the record and mechanisms of climate change on the planet, missed its orbital insertion and was lost. Post-mission analysis traced this failure to the use of English units of measurement in a command stream that required metric units. Appropriate derision of JPL and agency competence followed this jaw-dropping revelation. Next, the Mars Polar Lander stopped transmitting shortly before its entry into the martian atmosphere; the inference is that it had crashed on the martian surface. Serious soul-searching at the agency followed these failures, but only about the means, not about the ends. One great concern was with Goldin's alleged devotion to FBC (although it is hard to ascribe both of these lost missions to this paradigm, since their combined cost was over $300 million in FY1999 dollars) and not with the idea of a continuing series of Mars missions designed to follow the water. Following these failures, the revamping of the program now assured that the means of future missions to the red planet would each cost an appropriately staggering amount of money, all in pursuit of the elusive ends: water and therefore possibly past life.

The Human Space Program

Human spaceflight efforts following the demise of President George H. W. Bush's Space Exploration Initiative (SEI) in 1992 included the continuation of the space shuttle program, with its wide variety of satellite deliveries and life science experiments, as well as servicing missions to the Hubble Space Telescope. The goal of building a permanently occupied human space station had not been abandoned, but it had been reimagined. Space station *Freedom*, initially proposed by President Ronald Reagan in 1984, went through several design iterations, changes that delayed the start of its construction and increased the cost of the program. Despite program review after the *Challenger* accident and the grounding of the shuttle fleet by hydrogen leaks, NASA pressed on with station design and redesign. The fits and starts of the program led to exasperation in Congress, where it survived a 1993 vote in the House of Representatives by a margin of one. The human spaceflight program had reached a crisis of both confidence and capability.

Despite the decline and termination of the SEI, debate continued over the future direction of human spaceflight, as outlined by the Paine, Ride, and Augustine 1990 reports. At this time, one of the biggest concerns of science and technology policy was the problem of nonproliferation. The Soviet Union had dissolved, and there were concerns in the West that Russian scientists might sell their services and capabilities to rogue nations to make the infamous "weapons of mass destruction" and cause the spread of nuclear capabilities. It was thought by some that a joint space project involving both the United States and Russia in collaboration would keep the Russian military industrial complex safely occupied and under the scrutiny of its Western partner nations. The Soviets had built a fairly large and capable space station in the 1980s called Mir. Soviet cosmonauts conducted routine, extended stays on Mir, arriving and returning on their Soyuz spacecraft. Announced in 1993, the logical, initial starting point for this new East-West spirit of cooperation, called Shuttle-Mir, had rotating crews taking the space shuttle to Mir, where crews would live together, showing that we could work together peacefully in space.[9]

Between 1995 and 1998, there were eleven shuttle missions to Mir, where American astronauts spent close to a thousand days in orbit aboard the Russian space station. Joint operational and flight techniques were developed between the two countries. Despite some shaky moments (including a fire onboard the station requiring quick and decisive action by the crew), both parties considered this cooperative flight experience successful, leading to the final redesign of the U.S. space station as a new International Space Station (ISS).[10] This new design would be based on some key components provided by Russia. The *Zarya* (Functional Cargo Block), launched in 1998, and the *Zvezda* habitat and laboratory, launched in 2000, would become the nucleus of the new modular space station. Over the course of the next decade, twenty-seven shuttle flights and six Russian Proton and Soyuz flights would be needed to assemble the ISS in orbit. Starting in 2001, the ISS has been continuously occupied by crew, including during the period of the thirty months that the shuttle was grounded after the *Columbia* accident of 2003. With the delivery and attachment of the Alpha Magnetic Spectrometer, assembly of the ISS was finally completed in May 2011.

In the early years of the new millennium, as assembly of the ISS finally began, some in the agency considered the possible next steps for humans in space. Despite the failure of the SEI and the ongoing difficulties of the robotic Mars program, the obsession with human Mars missions was firmly entrenched within NASA. Most of the agency's advanced planning people spent their time devising new architectures designed to achieve that elusive goal. A core group of engineers in the Exploration Program Office at the Johnson Space Center continued to evaluate the requirements and difficulties of a human Mars mission, as well as alternative concepts involving return to the Moon. The post-SEI analysis of the Houston engineers had determined that with the launch of a few large expendable rockets and a couple of shuttle flights, we could return humans to the Moon.[11] Their analysis showed that the massive infrastructure creation outlined in the 90-Day Study was not strictly necessary, at least for the initial steps of human lunar return, especially if lunar resources (oxygen) were incorporated into the architecture. Such a mission would have limited stay time and capability, but at least it established a foothold on the lunar surface and could become a point from which the possibilities of extended presence could be investigated.

Studies of these architectures and plans continued, including investigations of missions to destinations other than the Moon. An early study mission favorite of NASA was the Lagrangian-point (L-point) mission, a human mission to one of the gravitational balance points in the Earth-Moon system, a point at which Earth and Moon appear to be stationary in the sky. The problem with L-point missions is that there is nothing there, except for what we put there. In the future, L-points could become critically important as staging areas for missions to the planets, or to collect exported material such as water launched from Earth or from the Moon. Although there was some interest in human missions to near-Earth asteroids, they were thought to be of much less importance, something to be reconsidered in the future. At the time, little was known about most of these objects, and asteroids had most of the disadvantages of a Mars mission (months of travel time, poor abort capability, and so on) with few of the benefits—for instance, most of these objects are simple, relatively homogeneous rocks, offering little in the way of exploratory variety.

As study of human Mars mission architectures continued, two things became increasingly clear. First, several technical developments, some of significant magnitude, were needed before human missions to the planet were feasible. Some of these involved "known unknowns," things we know that we need but don't yet have, like nuclear rocket propulsion or a solution to the dreaded entry, descent, and landing (EDL) problem,[12] while others consisted of the "unknown unknowns," problems of mission design or requirements that we don't even know about, let alone have any idea how to solve. Given the state of our knowledge, these studies showed that a human Mars mission is not possible in the near future. Moreover, even at favorable opportunities, a Mars mission requires between one and two million pounds to low Earth orbit, most of which is propellant. It was estimated that to assemble in orbit a Mars spacecraft able to conduct a single human mission would require between eight and ten launches of a Saturn V-class heavy lift launch vehicle. The entire manned Apollo mission series of 1968–72 launched ten Saturn V rockets. This means that a single human Mars mission would cost several tens of billions of dollars, even if such a heavy lift vehicle existed. Other, more innovative approaches would have to be considered.

A key step toward understanding how to conduct a human inter-planetary mission came in 1990 when Robert Zubrin, an engineer from Martin-Marietta, published his Mars Direct architecture.[13] Although this plan bypassed the Moon, its significance for lunar exploration derives from its reliance on in situ resource utilization (ISRU). By manufacturing pro-pellant on Mars for Earth-return—processing the carbon dioxide (CO_2) in the martian atmosphere into methane (CH_4) for propellant for the return trip—significant mass savings are realized, thus greatly reducing the initial mass required in LEO. In addition, the Mars Direct architecture separated cargo and crew. A nuclear power plant and the processing equip-ment needed to make methane propellant from the atmosphere would be delivered to the martian surface two years before the crew arrived. This approach introduced a safety factor, in that, if the atmosphere processing was less efficacious than believed, the crew would not be trapped on the surface of Mars without the fuel to get home because they would launch from Earth only after the return trip fuel had already been manufactured and stored on Mars. Despite these benefits, engineers from both NASA and the aerospace industry were slow to accept even the minimal risk introduced by the ISRU scheme proposed by Mars Direct. This ingrained resistance to ISRU carried over to architectures for lunar return as well.

Despite the innovative nature of some ideas in Mars Direct, a human Mars mission was still too high a fiscal and programmatic cliff to scale. Thus, for most of the 1990s, despite the presence of the alleged fossil forms in ALH84001 and Goldin's lobbying, the Mars program remained a series of scientific robotic probes "following the water" while consuming more and more of the planetary exploration budget.

The Loss of Shuttle *Columbia* and Its Aftermath

In the years after Lunar Prospector, but before the Vision for Space Exploration (ca. 1998–2004), several attempts were made to restart lunar exploration, at least in terms of a series of robotic flights to address some of the new and exciting findings and unknowns about the poles. The vociferous debate over the presence and extent of polar ice continued and it was clear that more and higher quality data were needed to resolve the issue of water ice. Earth-based radio telescopes were barely able to see into parts of the permanently shadowed polar areas of the Moon. Both the eighty-meter Deep Space Network Goldstone and the huge, three-hundred-meter Arecibo radio dish mapped the south pole of the Moon, looking for evidence for the presence of ice. The data were inconclusive, since diffuse backscatter obtained solely from zero phase (monostatic) radar, in which the same antenna sends and receives the pulses, cannot uniquely distinguish between rock and ice. The bistatic technique, where the receiving antenna is different and separated by a known distance from the transmitter, can uniquely determine this, providing evidence that caused numerous scientists to support the interpretation that ice had been detected.[14] As a believer in the polar ice hypothesis, I can attest to our desire to obtain new, high quality data from an orbiting radar experiment. The problem was finding a ride to the Moon. Japan has long harbored lunar dreams and had prepared a most ambitious orbiting mission, SELENE (later renamed Kaguya), a spacecraft the size of a school bus with a payload of almost every remote sensing instrument known to us.[15] But SELENE kept getting delayed, then was grounded by a launch vehicle failure and a Japanese economy in recession. Europe kept studying lunar missions, including both orbiters and a south polar lander, but each time such a flight was proposed, it was deferred. After downsizing their lunar mission into a small, technology demonstration, Europe's SMART-1 orbiter finally launched in late 2003, taking over a year to spiral out to the Moon using solar electric propulsion. The SMART-1 mission had limited instrumentation but it contributed to our knowledge of the poles by improving our mapping coverage and extending observation of polar lighting over a longer season.

A project sponsored by the Defense Advanced Research Projects Agency (DARPA) in 2003 looked at the possible impact of using lunar material resources to create new capabilities in space. This effort was mostly a paper study, although its authors hoped to parlay that report into a series of small robotic missions designed to follow up on the polar discoveries. I was working at the Johns Hopkins University Applied Physics Laboratory (APL), a university-based research organization similar to NASA JPL, when they studied that effort. We outlined concepts for a fleet of small satellites, each less than 100 kilograms, that could be operated in tandem to create high-resolution data on lunar polar environments and materials. Such a mission series would yield definitive answers for some polar questions, allowing us to understand if developing lunar water was feasible and what leverage in spacefaring capabilities it would yield. Although this topic is potentially the kind of transforming, "far out" idea DARPA claims to seek, the study was not approved to the next level of development, dashing the hopes of lunar enthusiasts yet again.

Despite its deferment, several positive results came from this effort. We understood how to configure a small mission that could get high quality data for the poles. A parametric study by a group at the Colorado School of Mines led by Mike Duke, former lunar sample curator and one of the masterminds of the 1980s lunar base movement, led us to understand the break points for lunar mining.[16] For example, what concentration levels of water make the effort of lunar mining economically worthwhile? It turns out that water concentrations of at least 1 weight percent are needed to balance the estimated costs of extraction, including the transportation system. Fortunately, we already knew that the existence of such quantities was likely: LP hydrogen data indicated an average concentration of 1.5 weight percent for the entire polar region, suggesting the possibility of even larger amounts of water in the shadowed areas.

On February 1, 2003, the space shuttle *Columbia* broke apart during reentry.[17] All seven crewmembers were killed. Until the cause of the accident could be determined and a fix applied, the shuttle would remain grounded. As with the loss of *Challenger*, the previous shuttle disaster in 1986, this accident once again focused the nation's attention on the meaning and purpose of our national human spaceflight program. But this time, it did more than that. Sean O'Keefe, the new NASA administrator who had succeeded Dan Goldin in 2001, had a reputation as a "green-eyeshade" guy. He had been recruited to solve the agency's considerable budgetary and accounting problems with the International Space Station project, which he did during his tenure of office. Profoundly shaken by the shuttle accident, O'Keefe decided that if humans were going to continue to risk their lives by going into space, there must be some great and meaningful purpose of national import to the trip.[18] O'Keefe was determined to find it.

As most attention was directed to the *Columbia* accident investigation, a simultaneous and largely unnoticed parallel effort was undertaken to review the purpose and objectives of human spaceflight.[19] It was recognized that future budgets for the civil space program were likely to be tightly constrained, so any possible plans must be constructed for an austere fiscal environment. Given these limitations, was there a way to revitalize the human spaceflight program, or had we reached the end of the trail?

Among those considering the next steps during this interruption in the human spaceflight program was Klaus Heiss, an economist who had conducted some of the early feasibility studies of the shuttle. He became convinced that a return to the Moon with the aim of learning how to establish permanence through the use of local resources could be achieved under current budgets, laying the groundwork for later, more ambitious space efforts. A friend of the Bush family, Klaus went directly to see the president with his idea, who passed it on to NASA for detailed technical study. At Headquarters, Associate Administrator for Human Spaceflight Bill Readdy and members his team undertook a feasibility study of Heiss's plan for the establishment of a base on the Moon.[20] The group continued to work on the problem of a return to the Moon for the next year and a half, coming up with an approach that was both affordable and technically robust.[21] This "Gold Team" undertook an examination of the problem of trans-LEO human spaceflight, independent of previous advanced study work.

Over the remainder of 2003, a major cabinet-level study of the human spaceflight program was completed. The White House Office of Science and Technology Policy (OSTP), Office of Management and Budget (OMB), the National Security Council (NSC) and NASA all participated in this top-level review. Presidential Science Advisor John Marburger took an unusual and impressively independent path. Rather than rehashing the plans of previous "visionary" efforts and reports, he posed a fundamental question about human spaceflight: Why? What is our long-range purpose in space?

Many have wrestled with the "Why?" question over the years. Typically, this sort of pondering begins and ends in one's own subdiscipline within the space business. For most of the scientific community, the answer has been to study the universe. For aerospace industrialists, it is to get long-term government contracts to build the biggest, most expensive machines ever imagined. For agency bureaucrats, it is to start, expand, and manage a large continuing program. Marburger reexamined the issue and posed a question: Why not bring space into our economic sphere? For years we have heard about "limits of growth" and various environmental crises, ironically enough, much of this talk spurred by the pictures of Earth taken from the Moon by the Apollo astronauts. But space contains virtually unlimited quantities of material and energy, and thus, in theory, unlimited wealth. Why not focus on developing the technology needed to harvest that wealth for the benefit of humanity?

While the public focused on the *Columbia* accident investigation, two competing streams of thought emerged.[22] One, favored by OSTP and

OMB, focused on the practical and economic aspects of the space program. Could we reorient and retool the program to become a creator, rather than a consumer, of wealth? To do that, we would need to learn the techniques of planetary resource utilization, habitation, and extended operations. During Apollo, we had visited the Moon briefly for the purpose of scientific study and exploration. To extract useful products from the materials found in space, we would need an extended presence and different types of equipment and operations. Given the new findings about the nature and potential of the lunar poles, the Moon quickly emerged as the initial destination for the civil space program beyond LEO.

The second stream of thought about future directions was a very familiar one to longtime space observers: a human mission to Mars, the project that many had long dreamed about. Once again, NASA hoped to emerge from the ashes of *Columbia*, Phoenix-like, to take humankind to the planets. It is fair to say that not all at the agency were on board with this direction—Readdy's work on the lunar base studies, for example, showed considerable support for a return to the Moon—but it is equally fair to say that many were Mars-oriented, especially those involved in advanced planning decisions. A look through all the documents the agency produces detailing future missions shows that they largely revolve around the future, imagined needs of a humans-to-Mars program. Zubrin's influence had infiltrated the agency enough to incorporate some of the features of his Mars Direct architecture, including ideas like splitting cargo and crew mission segments and ISRU propellant production. But no matter which way it was cut, a human mission to Mars was still too big a stretch, a much larger effort programmatically, technically, and fiscally than returning to the Moon. The battle lines were being drawn.

These behind-the-scenes events were largely unknown to me as 2003 wore on. Then a chance meeting occurred at a November gathering of lunar base advocates in Hawaii. At that gathering, I described results for the lighting conditions at the lunar poles and the evidence for water in the dark areas. Also in attendance was Indian scientist Narendra Bhandari, who described his country's plans to fly its first mission to the Moon, the orbiter Chandrayaan-1. This small satellite was about the same size and capability as Clementine, and I felt an immediate affinity for their effort. During a break in the meeting, I approached Bhandari and asked him if they had considered flying imaging radar as part of the payload to map the dark areas near the poles. He replied that they had considered one, but imaging radars were too heavy and power-hungry and as such would not fit on a small spacecraft. I told Bhandari about our efforts to miniaturize a radar instrument for this purpose; we believed that we could make an imaging radar that would be less than 10 kilograms in mass and would use only 100 watts of power, an order of magnitude less than typical radar instruments. He promised to report our discussion to the Indian space agency and get back to me.

As the year waned, excited rumors circulated throughout the space community that a big announcement on space policy was imminent. The initial rumor held that President George W. Bush would unveil a new major space initiative in December, on the hundredth anniversary of the Wright Brothers' first flight at Kitty Hawk.[23] But that anniversary and celebration came and went without any announcement, causing some to believe that the policy plan was in trouble, when in fact it was merely in its final stages of review and briefing to Congressional and Executive personnel and staff. At a White House meeting in mid-December 2003, a final review of the new initiative was held. The idea was to announce the policy goal and then implement it, giving NASA a one-time budget augmentation of about $1 billion spread over the coming five years with the agency's budget rising only with inflation in subsequent years. Because the shuttle was an expensive, labor-intensive vehicle, its operating costs constituted a large fraction of the total NASA budget. Results coming out of the Columbia Accident Investigation Board urged that the shuttle be retired. The new initiative slated the shuttle for retirement, to be replaced by a new, less expensive human space vehicle, the Crew Exploration Vehicle or CEV, with both form and specifications to be determined. This replacement spacecraft would consume less of the annual agency budget, creating a "wedge" of money saved from the shuttle program that could be spent on missions beyond low Earth orbit. Thus, from the very beginning of the new initiative, the agency was being challenged to approach the effort in a new and innovative way. This was not to be a typical new program, with automatic "plus-ups" to swell the budget, so much as a new strategic direction. Within broad boundaries, the agency was given latitude to pursue its new destination goals in the manner that it perceived best. But go where?

During the final review meeting, President Bush was presented with summary arguments for programs that focused on lunar return and human Mars missions. He recognized that the real objective was to create a new and bigger pie, not to simply decide how to cut and subdivide the remaining pieces of an existing, dwindling, small one. With its proximity and known resources, the Moon offered the possibility of early accomplishment, but more important, it offered a way to make an eventual Mars mission easier and more affordable. Thus, the president decided that the objectives were to be both Moon and Mars.[24] In this sense, he was reinstating the goals of the abandoned Space Exploration Initiative that his father had proposed fifteen years previously. But this return to the Moon was different. In his major policy speech on the new Vision for Space Exploration (VSE), President Bush outlined the activities to be done on the Moon: to go there with the goals of staying for increasing periods of time to learn how to make useful products from what we found there. In other words, this lunar return was focused on sustained presence and the creation of new spaceflight capabilities.

Sustainability and the creation of new capabilities from what we find in space: these startlingly radical aspects of the new program were largely dismissed or ignored by many observers, who then went on to characterize the return to the Moon as merely the prelude to a human Mars mission. This false interpretation of the purpose behind the new policy was widespread within the agency, as well as in the space community as a whole. The confusion led to immediate and significant problems for many early strategic decisions on implementing the VSE. But when President Bush announced the new Vision on January 14, 2004, in a special speech at NASA Headquarters,[25] space advocates were encouraged. Finally, a coherent direction had been imposed on what was widely perceived as a foundering, directionless program. Despite the expected carping from some in the space science sector, most agreed that the new strategic redirection of the civil space program was worthy. The president announced that he was forming a commission chaired by former Secretary of the Air Force Pete Aldridge to examine ways to implement the new Vision. This commission was to report back to him in six months. To my surprise, about two weeks after the announcement, I received a call from the White House asking me to serve on this commission, an assignment that I was more than happy to accept.

NASA now had a new direction and the possibility of a fresh start after the *Columbia* disaster. For lunar scientists and advocates, the new VSE was an intoxicating promise to revisit our object of desire, and to develop new technologies that would enable long-term human presence off-planet. For the Mars community, there was subdued rejoicing and a somewhat irritated acceptance of being relegated to a "long-term" objective. But as things progressed, we soon discovered that despite the clear strategic direction the VSE provided, following it was not going to be so easy.

5

Implementing the Vision

On January 14, 2004, President George W. Bush unveiled the Vision for Space Exploration (VSE) during a visit to NASA Headquarters. The product of almost a full year's review by the White House and NASA, it outlined a strategic path to reestablish a sense of purpose and direction for the nation's civil space program.[1] The VSE consisted of four major elements: return the Shuttle to flight to complete construction of the International Space Station; develop and build a new human spacecraft, the Crew Exploration Vehicle (CEV), and eventually retire the shuttle; set our space program on a course to the Moon with the object of "living and working there for increasing periods of time"; and eventually, undertake a human mission to Mars. No deadlines were declared for these milestones, although lunar return was given the scheduling guideline of "as early as 2015 but no later than 2020." Budgetary direction to support the new Vision was articulated: a single, one-time budget augmentation of about $1 billion spread over five years, with the remainder of the funding needed for the implementation of the VSE, about $11 billion, freed up from existing funding through the retirement of the shuttle, since servicing and preparing it for flight was a labor-intensive activity that consumed a large fraction of the agency's operational human spaceflight budget.

The announcement of the new VSE caught many off guard, and although its reception was mixed in some quarters, the overall reaction to it from most space program observers appeared to be positive. As part of the roll-out, it was announced that a commission headed by former Secretary of the Air Force, Edward "Pete" Aldridge, would meet to study how to implement the new space vision and report the various options to the White House within 180 days.

The Aldridge Commission met for the first time in early February 2004 in an office complex in Arlington, Virginia. I was a staff scientist at the Johns Hopkins University Applied Physics Laboratory at that time. The other commissioners were planetary scientists Laurie Leshin and Maria Zuber, astronomer Neil DeGrasse Tyson, former Congressman Bob Walker, General Les Lyles, former Deputy Secretary of Transportation Michael Jackson, and Hewlett-Packard CEO Carly Fiorina. Most of us had served on various space advisory committees before and knew what was expected of us. At our first meeting, we went around the table, assessing exactly where we all stood in regard to our task and the Vision. While all were supportive and excited about the new VSE, everyone expressed a concern for the need to develop and to articulate a strong rationale for a continuing, sustained program. Perhaps with a bit too much optimism, we thought that we could weave such a rationale into the report, both as an underpinning logic to our recommendations (whatever they were to be) but also for use by both the administration and NASA to help "sell" the VSE to congressional appropriators.

Our work was prefaced with a series of presentations given by the various administrative codes of NASA (Space Science, Human Spaceflight, and so on). These summaries were designed to inform our group on what the various parts of the agency saw as their mission challenges and what they planned to do in response. Ed Weiler, then associate administrator for space science, gave one of the earliest presentations. A bullet point on one of his slides read, "activities on the Moon will be minimized and restricted only to those that support human Mars missions." It was surprising to see this statement; after all, President Bush's speech had been specific and concrete about activities for the Moon (something unusual in a presidential speech), indicating that learning to live and work on the Moon for increasing periods was a major objective of the VSE. It became clear that Weiler's understanding, along with several others at the agency (especially in the Space Science section) was something entirely different from the straightforward language of the Vision. He took the position that the VSE was almost entirely about human Mars missions. When I pointed out this discrepancy to Weiler, my concern was dismissed without a meaningful response, only a comment to the effect that "this is how *we* understand the Vision." Almost immediately, the lunar parts of the VSE were deemphasized, with a gradual yet perceptible shift towards using the Moon as a mere testing ground for the Mars mission.[2]

I was able to trace this evolution in a series of both official and internal agency documents, ending with the Weiler interpretation of "minimal activity on the Moon." The origins of this intellectual disconnect go back to the late 1990s and early 2000s, with something called the NASA Decadal Planning Team—DPT, later called NEXT, for NASA Exploration Team.[3] Chartered under Administrator Dan Goldin, this group was tasked with mapping a path for human missions beyond low Earth orbit, ultimately leading to a Mars mission. Although Goldin was fixated on Mars, the DPT focused primarily on nearer term objectives in cislunar space, including the libration points—that is, points in space that remain fixed in relation to Earth, Moon, and Sun—and the lunar surface. Additionally, both asteroid missions and missions to the moons of Mars were considered. The rationale behind the stepping-stone approach was to offer flexibility to some future administration that might be interested in a long-term, deep space goal for human spaceflight. The agency still maintained their fixation with Mars, and the idea was prevalent throughout the DPT that any activity on the lunar surface detracted from and delayed the Mars mission.[4] In consequence, mission planning before the VSE did consider lunar missions but only in a cursory manner.

Although we knew by 2000 that the poles of the Moon probably harbored ice deposits, water production from polar deposits was not included as part of any lunar surface architecture. Moreover, the key finding from

Clementine that areas of quasi-permanent sunlight could be found near the poles (enabling sustained permanent presence on the Moon) was acknowledged but not integrated into a useful surface operations plan. In fairness, during those years of the DPT-NEXT, the agency had no authorization to proceed beyond the ISS and shuttle, so their studies, while interesting and useful in terms of what capabilities could be developed, could not be implemented or even integrated into any long-range strategic plan.

The thrust of agency efforts in this era was the emphasis on the quest for extraterrestrial life,[5] both the actual search for life elsewhere and use of "the quest" as a driving political and programmatic rationale for exploration beyond LEO. In part, this was a natural outgrowth of the parallel and continuing robotic Mars program. However, that intellectual milieu meant that when human missions were to be considered, lunar surface activities (which were thought then to be largely irrelevant to studies of life's origins, an incorrect but widely held belief) tended to be deemphasized. I believe that this fixation with finding life on Mars held and still holds the agency hostage, unable to consider activities on the Moon to be anything other than a technology demonstration in preparation for a human Mars mission. Thus, when the VSE was announced, although considerable verbiage was devoted to detailing the activities to be undertaken on the Moon, the agency heard only one word as its destination: Mars. In consequence, the previous focus on the "search for life" carried over to become the underpinning science rationale for the new VSE.

The Vision, as originally articulated, was specific and quite different. The Moon in the VSE was to serve as a laboratory, a workshop, and a logistics depot. The idea was to learn how to use the material and energy resources of space (including lunar polar ice) to create new spaceflight capability.[6] Many misunderstood or dismissed this latter concept. The Moon's role in the VSE was mischaracterized as "landing the Mars spacecraft on the Moon for testing and refueling." In fact, the significance of the Moon in the Vision was to use it to develop technologies useful for future missions, as well as to develop lunar resources to fuel the missions to distant destinations. This concept *was the vision* in the VSE.

The Aldridge Commission held public meetings in Dayton, Atlanta, San Francisco, New York, and Washington, D.C., to gather information and testimony from local experts and to give the public a chance to weigh in with their concerns and hopes for the direction of human spaceflight. Subgroups of the commission undertook fact-finding trips to the various NASA field centers with the aim of understanding whether all of the centers were needed to execute the VSE, or whether a different management model might be employed to make NASA more efficient. To evaluate the possible roles of the commercial sector in implementing the VSE, we gathered a considerable amount of information from the space industry. All the while, I attempted to revector the effort back toward its original intent of learning how to use space resources.[7] Finally, we examined the configuration of management within NASA and deliberated on how to make the agency both more efficient and more accountable in the completion of its assigned tasks.

The Aldridge Commission report was issued in July 2004.[8] Even though its recommendations were reasonable and moderate, only a few were seriously considered and even fewer were eventually implemented. Our idea for NASA to procure delivery of goods and people to low Earth orbit eventually resulted in the Commercial Cargo and Crew program. Engineering management buzzwords like "spiral development" were eagerly embraced by the agency, but such enthusiasm did not move the ball forward to any great degree. Some ideas were conspicuously ignored, such as resurrecting the National Space Council to act as an oversight body for NASA and the idea to turn field centers into federally funded research and development centers (FFRDC), a mode of operation in which a university manages an agency field center—NASA's Jet Propulsion Laboratory, managed by Caltech, operates this way. This structure permits easier personnel recruitment and turnover, and it allows centers to seek new business from the private sector—features designed to keep field centers technically strong and their management more nimble and responsive to rapidly changing fiscal and programmatic conditions.

That the commission's report was largely tabled is probably not too surprising. However, I *was* surprised at what I perceived to be the extreme inertia of the agency in getting the VSE started. The obvious first step in any lunar return was to fly a robotic mission to follow up on the Clementine and Lunar Prospector polar discoveries. Mapping the Moon globally at high precision and resolution would create a database of strategic knowledge to help plan and execute future missions. An agency call went out to the scientific community (called an "Announcement of Opportunity," or AO) to propose instruments to fly to the Moon on a mission called the Lunar Reconnaissance Orbiter (LRO). Among the many specifications of new, required strategic data was one to "identify putative deposits of appreciable near-surface water ice in polar cold traps at ~100 m spatial resolution."[9] At the Johns Hopkins University Applied Physics Laboratory, I was part of a team that proposed an imaging radar for the LRO mission to address this requirement.[10] We also proposed flying a smaller, less capable radar instrument that could fit on the Indian Space Research Organization's forthcoming Chandrayaan-1 mission to the Moon as a guest payload. Radar would be useful to map the shadowed, cratered regions near the poles, data needed to study the RF reflection properties of the interiors of these craters to determine if ice might be present there.

To our surprise, the radar instrument (Mini-SAR) was selected for India's Chandrayaan, along with a spectral imager (Moon Mineralogy Mapper or M^3) as a second American guest payload—but not for America's LRO mission. In fact, the selected payload for LRO contained *no* radar instrument at all. Instead, to infer the distribution of water, a Russian neutron detector was chosen, a design that experts told us was probably inadequate to produce hydrogen maps of the poles at the high resolution required by the AO. These decisions, made in early 2005, caused great concern among those of us working toward lunar permanence and resource use; it appeared to be a selection designed more to check off a box on a chart rather than one geared toward the gathering useful strategic knowledge. Our rejection was appealed to the Administrator of NASA and after some wrangling, the Mini-RF radar was approved for flight on LRO. This administrative fracas led to some resentment toward the radar experiment by some of the

LRO project people at NASA–Goddard Space Flight Center. Our Mini-RF experiment was accommodated on the LRO mission as a "tech demo," and although the project had been directed by senior management to accommodate Mini-RF, our team had to fight for observing time and spacecraft resources during the nominal mission.

Flying the radar on the Indian mission was more gratifying.[11] Chandrayaan-1 was India's first mission to deep space and the Indians were quite excited and proud of their maiden efforts in trans-LEO spaceflight. The Chandrayaan spacecraft was relatively small, about the size of Clementine, yet very capable. It carried not only precision imaging cameras but also flew instruments to map the mineralogy and chemistry of the surface. The two American experiments flown on Chandrayaan, our Mini-SAR radar instrument (built by Raytheon and APL) and the Moon Mineralogy Mapper (built by the NASA Jet Propulsion Laboratory) had to get approval from the State Department before we could fly them to the Moon. I was told at the beginning of this effort that because of sensitivities to export control issues, it was highly unlikely that we could get permission to fly the Mini-SAR on India's mission. But it turned out that our application to participate on Chandrayaan coincided with a presidential-level initiative to improve US-India relations. As a result, the State Department was very supportive of our effort. A last-minute intervention by the White House led to the approval of the export license. Mini-SAR became the first American scientific experiment to propose and be selected to fly on an Indian space mission.

I made almost a dozen trips to India over four years. Each one-way journey required roughly twenty-six hours in transit and lasted only a few days; the nearly twelve-hour time difference between India and America played havoc with my internal clock. The upside was that the Indians were a pleasure to work with. They were enthusiastic about going to the Moon and their mission received a lot of local publicity. Whenever I told anyone why I had come to India, the universal response was excitement and an eagerness to learn more about the mission. After selection, the actual work of flying an experiment in space largely involves attendance at innumerable meetings, where the arcane details of each system and every part are described and debated. During design, assembly, and test, scientists have little real work to do; we determine and define the parameters of the instrument and devise a plan to collect the data, but ordinarily, our work happens during and after the flight, when the data streams down and must be reduced, formatted, and interpreted. Many of us view preflight work as paying dues for the fun work to follow. And as many can attest, all of this planning and effort can just as easily go up in flames if the launch does likewise.

The LRO version of the instrument was a bit more challenging. The LRO radar was to operate in two radio frequencies at two different ground resolutions, but the Mini-SAR and Mini-RF instruments were basically the same. In terms of operation time, Chandrayaan was scheduled to launch about a year before LRO. It was hoped that we would obtain full data for both poles from Chandrayaan, which in turn would help us plan to take high-resolution data of interesting areas with the LRO Mini-RF build. During an extended mission and with a little luck, we might even be able to collect enough data to make a radar reflectance map of the entire Moon, detailing slope distributions and locating jagged rock fields on a global basis.

The Fate of the VSE at NASA: What's the Mission?

Although much of my time was spent working on the two radar instruments, I was also on a number of advisory and analysis groups at NASA dealing with the implementation of the lunar phase of the VSE. Once the Aldridge Commission submitted its report, the new Exploration Systems Mission Division (ESMD) began its process to define the spacecraft, dubbed Project Constellation,[12] and the missions that would constitute our nation's new space program. Despite the clear, strategic direction NASA had been given regarding the Vision, in those early planning stages, there was growing cause for concern about the fate of the VSE.

The head of ESMD at NASA, Admiral Craig Steidle, who came to the VSE from another large engineering project, the Defense Department's Joint Strike Fighter program, had no spaceflight experience. The then-current vogue in large engineering projects was a technique called spiral development.[13] The spiral plan called for four sequential stages: develop requirements, analyze risks, build and test, and evaluate results. The product becomes the new "block" to be refined in the next spiral. Another name for this process is "build a little, test a little." The idea is to pursue the most promising designs by not committing to a final version until significant experience and test data are acquired. NASA's devotion to this new management voodoo was reminiscent of many previously embraced business school fads, such as Total Quality Management (TQM). One notable employer of spiral development, amazingly enough, was the F-35 Joint Strike Fighter program, the project whence Steidle came and one renowned for being years behind schedule and billions of dollars over cost.

Thus, from the first step, with the development of requirements and a seemingly endless exercise called technology "road mapping," the VSE at NASA started off slowly—and then tapered off. Many different experts in science and engineering were brought together at great time and expense to opine on what the new spaceflight systems had to accomplish, in what order, and to what degree of fidelity. This involved building complicated spreadsheets whose contents were populated by technologies, instruments, and knowledge needs. The problem with this activity is that too often, problem definition becomes a substitute for actual programmatic progress, since critical decisions can always be deferred while awaiting better defined or more perfectly understood requirements.

Lest it seem that no progress was being made, there was one activity in the post-VSE announcement era that warrants special mention. Associate Administrator for Spaceflight Bill Readdy had pulled together an informal study team (taken from his section of engineers and experts) to examine a possible path to implementing the VSE. While the agency had an internal "red team/blue team" study effort, which came up with the lunar "touch-and-go" concept, Readdy put together what was called the "Gold Team," whose mandate was to examine unorthodox approaches to implement the Vision. The Gold Team looked at the issue of developing a new human trans-LEO capability, while at the same time returning the shuttle to flight and completing the construction of the ISS.

The Gold Team found that the original charge to the agency—to return to flight, finish building the ISS, and develop a new human space vehicle, all with the aim of returning to the Moon by 2015—was achievable if certain architectural choices were made *early*. The most significant feature of their approach was to retain the shuttle launch infrastructure to support the first two milestones and then use that asset to build the shuttle side-mount heavy lift launcher, a derived vehicle that used shuttle engines, external tank, solid rocket boosters, and all of the existing Cape infrastructure. The advantage of shuttle side-mount was that by using existing pieces, it would require minimal new development. As will become clear, the reason that Project Constellation was cancelled is rooted in escalating, higher than expected early development costs that continually pushed its projected first flight farther and farther out into the future. If NASA had chosen to go down the path of the Gold Team, we would have completed the ISS and retired the shuttle on schedule, and the new shuttle side-mount would have been ready to fly humans by 2015.

The advantage of the Gold Team approach was that by adopting shuttle side-mount, most of the development costs for new deep-space systems could be focused where they were most needed: on the new CEV and a robust program of robotic precursor missions to the Moon. The CEV at this stage was undefined; it could have taken the shape of an Apollo-type capsule, as it ultimately did under the Constellation program as the Orion spacecraft, or it could have been the more flexible "bent biconic" design,[14] an aerodynamically shaped body similar to that of the Blue Origin commercial spacecraft. This latter design could have served as a pathfinder development for a Mars entry vehicle, as they have similar aerodynamic shapes and would be able to land on its tail under thrust, permitting soft, dry landing at the launch site, like the shuttle. Separate crew modules derived from ISS hardware would serve as cislunar transfer vehicles.

The original VSE called for a significant and robust program of robotic missions, but the Gold Team took this further by using such missions to emplace infrastructure on the Moon. A large robotic lander was planned, designed to use solar-electric propulsion (SEP) and large solar arrays to spiral out slowly from LEO to the Moon and then use a LOX-hydrogen rocket to land up to several metric tons on the lunar surface. After landing this payload, a mobile lander platform would separate and the large solar arrays that powered the SEP would become part of the electrical power-generating infrastructure of the outpost. Through this approach, we would begin to establish a permanent lunar surface outpost, a facility eventually to be used by humans. By predeploying habitats and subsystems on the Moon using unmanned spacecraft, we could make the human-rated systems smaller (reducing development costs), yet adequate (taking advantage of preemplaced assets). The innovative use of robotic missions by the Gold Team was a significant departure from ordinary agency practice, whereby robotic missions are used primarily for the acquisition of scientific and engineering data, which are then used to design the human vehicles. Instead, the Gold Team advocated using robotic assets in tandem, and in parallel, with the human spacecraft and missions.

Although the Gold Team architectural approach had much to commend it, both in technical and in fiscal terms, Readdy was not the agency point man designated to make these choices. Steidle and the Office of Exploration were aware of this work but did not take it seriously, insisting instead on pursuing their road mapping and spiral development approach—which, in this case, consisted mostly of deferring decisions indefinitely. The only effort proceeding to actual flight was LRO—planned as the first in a series of robotic exploration precursor missions sent by spacefaring nations around the world to the Moon.

A New Administrator and the ESAS

Early in 2005, Sean O'Keefe announced his decision to leave NASA to become chancellor of Louisiana State University. Michael D. Griffin was tapped as the new administrator, coming to NASA with an impressive background of engineering and management experience backed up by seven university degrees.[15] I knew Mike from the Synthesis Group days, when he was one of our senior members, and from the Clementine project, where he was deputy director for technology in the Strategic Defense

Initiative Organization. Griffin also served as the associate administrator for exploration at NASA during the SEI days, although as we have seen, that program was abandoned. A visionary, Mike was and is a strong advocate for a vigorous and expansive human space program. Around the time of the VSE rollout, Griffin had led a study sponsored by the Planetary Society, outlining an architecture for human missions beyond LEO, primarily driven by the requirements for human Mars missions.[16] This plan was notable for its use of a crew launch vehicle derived from a single shuttle solid rocket booster, an innovation that generated much comment and subsequent controversy.

Griffin decided that NASA had wasted the last eighteen months with road mapping exercises and spiral development and summarily dismissed Steidle. In his place, Griffin brought in Scott "Doc" Horowitz, a former astronaut and the engineer who had come up with the idea for the "stick," the SRB-based launch vehicle. To move the ball down the field, one of the first things Griffin did after assuming agency leadership was to convene an ad hoc study group to design an architecture for missions beyond LEO. This effort, dubbed the Exploration Systems Architecture Study (ESAS),[17] was conducted from midsummer to the fall of 2005. The lead engineer was Doug Stanley of Georgia Tech who led a team of mostly NASA engineers from Headquarters and the field centers. I was a member of this group, but my involvement was focused only on lunar surface activities and the identification of possible landing sites. I was not involved in any major decisions about the spacecraft and launch vehicles of the architecture.

The ESAS team began with a set of assumptions about the requirements of the new transportation system and how it would be used. The study embraced the recommendation of the Columbia Accident Investigation Board (CAIB)[18] to separate crew and cargo, thought to be a safety issue, although no one could really give a logical rationale for it. There was a sense that launching a rocket with the crew positioned on the side of the vehicle, like the shuttle, was inherently unsafe, although this specific idea is not part of the CAIB report. It is difficult to justify this edict on technical grounds, since 134 shuttle flights safely launched people in this configuration and in the one accident that occurred during launch, *Challenger*, the crew and their cabin survived the explosion and would have lived had the cabin been equipped with parachutes; they were instead killed on impact with the sea. This ground rule was important because it meant that adoption of a shuttle side-mount design as a launch vehicle would likely require three vehicles per lunar mission rather than two, a consequence later used to justify the elimination of the side-mount option. The new architecture was mandated to serve ISS crew and cargo requirements, in addition to lunar surface missions, even though the then-current plan called for ending American participation in the ISS around the time that the new systems were to come online. Certainly this was not the first time in the history of the space program that an architecture was devised under the constraints of arbitrary and illogical ground rules, but serious consequences were to emerge from these boundary conditions.

The ESAS work came up with an interesting solution to the architectural problem of launch, something they called the "1.5 launch vehicle" solution. In brief, the study advocated the development of two different launch vehicles: a smaller (20 ton) crew launch vehicle identical to the Planetary Society's SRB "stick" rocket (Ares I), and a larger (130 ton) shuttle-derived, inline vehicle (Ares V) to carry cargo and heavy payloads (launching two differently sized vehicles led to the nickname). A single mission would use both vehicles; the lunar lander and Earth departure stage would be launched on the large Ares V as "cargo," while the crew would be launched separately on the smaller Ares I. The two spacecraft would rendezvous in Earth orbit, dock, and then depart for the Moon. The rest of the mission profile followed the same pattern as the Apollo missions: lunar orbit insertion, landing, ascent, rendezvous, and return to Earth in the Orion CEV. Both the Orion CEV and the Altair lunar lander were larger, more capable versions of the Apollo CSM and Lunar Module. Because of Constellation's similarity of appearance and mission profile to the Apollo missions, Mike Griffin once referred to this architecture as "Apollo on steroids," an unfortunate characterization that reverberates to this day.

The ESAS report was released in October 2005 to less than universal acclaim, with many noting the similarity of the new plan to the old Apollo template. In fact, although the new plan would create considerable capability, the flying of individual, one-off missions whereby most pieces are discarded after a single use, reverted us back to an earlier era of the space program. As in Apollo, only the crew command module (Orion) would return to Earth. The individual missions would carry a larger crew of four and stay on the lunar surface longer, up to two weeks. The large capacity of the Altair lunar lander meant that significant cargo could be placed on the Moon, permitting an outpost to be established with a minimal number of launches.

An important point to understand about the ESAS architecture is that its heavy lift launch vehicle (Ares V, starting out at 130 metric tons, but expandable to 160 metric tons) is scaled for human Mars missions staged entirely from Earth; its utility for the lunar missions is genuine but only incidental. A mission to the Moon requires roughly 100–120 metric tons in LEO (depending on how the mission is configured, its equipment and destination). This could be accomplished with two medium-class heavy lift launches (70 metric ton; shuttle side-mount) or the launch of a single, large vehicle (Saturn V-class). The Ares V is much larger than what is needed for routine missions to the Moon. But if the requirement is to deliver pieces of a 500-metric-ton Mars spacecraft to LEO, then transporting it with as few launches as possible greatly reduces overall risk. Clearly, the ESAS was looking ahead to the future where it was thought that NASA would get only one chance to develop an entirely new space transportation system in the new century and its objective was to plant the American flag on Mars. The seed of the problem had been planted and the future of spaceflight envisioned through the lens of the Apollo program—with disposable spacecraft and everything launched from Earth—became unaffordable and thus unsustainable. It still is. More important, it discarded the original point of the VSE: to learn how to use the material and energy resources of the Moon to create *new* spaceflight capability.

Lest anyone think that this latter point was unclear or had been inadequately presented—after all, the VSE had been unveiled in a relatively brief presidential speech two years previously—in March 2006, Presidential Science Advisor John Marburger gave one of the finest speeches I ever heard on the meaning of the VSE and on a rationale for spaceflight in general.[19] Speaking at the annual Goddard Space Symposium, Marburger carefully laid out the physical difficulties of spaceflight and articulated why the Moon has a critical role to play in creating new capabilities in space. He posed a key question: "What is the purpose of our civil space program?" Marburger then stated that "questions about the vision boil down to whether we want to incorporate the solar system in our economic sphere, or not." And he provided an answer: "For a space program to serve national scientific, economic and security interests, we must learn to use what we find in space to create new capabilities, starting with the material and energy resources of the Moon." Marburger also pointed out that such a mission had much greater long-term societal value than space activities "confined to a single nearby destination or to a fleeting dash to plant a flag." Because the Moon is close, reachable, and useful, it was chosen as the centerpiece of the VSE. Mars was a destination reserved for the future, after we had mastered the new skills and technology needed for spacefaring.

Apparently, few in the agency heard or read Marburger's speech because NASA either misunderstood their charter in the VSE or deliberately torqued it away from the intended direction. An Exploration Strategy Workshop, held in April 2006 in Washington, gathered an international cadre of about 150 space experts for a four-day meeting to identify why we were going to the Moon and how to best accomplish those goals. The boundary conditions were the features and limitations of the ESAS architecture; otherwise, the agenda was completely open. I was stunned by the premise of this meeting. The VSE speech of January 2004 was the clearest, most unambiguous strategic direction given to the space agency by a president since John F. Kennedy's Apollo declaration.[20] Yet two years later, NASA decided to convene a group to come up with a rationale for lunar return and to envision a set of activities once we got there. The agenda for and mindset of this meeting convinced me that the VSE was in serious trouble.

The workshop attendees deliberated over the course of three days, drawing up six major "themes" for lunar return: human civilization, scientific knowledge, exploration preparation, global partnerships, economic expansion, and public engagement. Flowing from those six broad-based themes was a "grid" of specific requirements and activities, 186 entries identifying what would become the input to the succeeding Lunar Architecture Team (LAT). Although the ESAS specified the hardware and mission profiles, exactly how they would be used, which events and in what order they would take place on the Moon, were yet to be specified. With workshop results having "told us" why we were going to the Moon, we could begin to focus on the "how."

Among the topics to be grappled with were the sites on the Moon to be visited, whether to set up an outpost or conduct multiple sortie visits, and which investigations to conduct and in what order. The LAT consisted of scientists and engineers who would meet several times per year but would mostly perform their work at their home institutions. Tony Lavoie, an engineer from NASA-Marshall in Huntsville, chaired the first LAT. Tony and I had previously worked together on planning the later-canceled second lunar robotic mission, a lander and rover designed to map and characterize the ice deposits in the permanently dark areas near the poles. We knew something about the lunar polar environment from Clementine and LP, and the soon-to-fly LRO would add detailed knowledge that would allow us to pick the optimum landing sites for surface activities.

The first LAT came up with solid, defensible conclusions, especially in regard to mission mode and priority activities.[21] The most important decision made was to focus lunar return on the establishment of an outpost near one of the poles of the Moon; *which* pole was to be decided after LRO and some surface rover data had been collected. The principal reason for an outpost is that you can concentrate assets at a single locality and rapidly build up capability. The alternate approach is to conduct sortie missions, which permit visiting many different sites with wide geographic and geologic diversity but preclude the concentration of assets, since the sites would be abandoned after each mission. The sortie strategy was the Apollo template writ large; the outpost approach would mean permanence, or at least long-term habitation, and the opportunity to build a production-level resource processing facility. Unlike many within NASA, Lavoie clearly understood the real meaning of the VSE: to return to the Moon and learn the skills needed for extended space presence and capability. Under his leadership, for the first time since its announcement, the VSE began to move toward a mission more inline with its original intent.

Despite its many good deeds, the LAT activity was still entrained within the NASA system and hence, was required to address the 186 entry "spread-sheet of death," as we called the table of activities and events to be accommodated while on the Moon. The practical effect of this was to diffuse the LAT effort away from its primary mission direction—a resource-processing outpost—into a nebulous, NASA lunar exploration mission. Most insidiously, the lunar "touch-and-go" on the way to Mars, the "real" objective, slowly crept back into the architecture. This happened largely during the second round of architectural planning (imaginatively named "LAT-2") in which sortie missions became the new baseline. In part, this was an agency reaction to an outcry during the public rollout of the LAT-1 plans in December 2006.[22] The usual suspects—the Planetary Society, media, various individuals—were greatly concerned that a significant amount of time and effort was to be expended on the Moon, thus delaying their Apollo-type "sprint" to Mars. A common phrase during this time was the expression of desire to get to Mars "in *my* lifetime," a requirement not derived from any programmatic principle I can discover. In short, the "sprint to Mars" cabal within and outside of the agency had struck back.

The report of the LAT-1 team at the end of 2006 was the high-water mark of the VSE. Although many in the agency still refused to "understand" precisely why we were going to the Moon, a solid, logical plan of action had been developed. Both the Chandrayaan-1 and LRO mission developments were proceeding well, as was our work on building the Mini-RF imaging radars to map the poles. Because LRO had grown in mass and had outgrown its original Delta II launch vehicle, a new Atlas booster with extra payload capacity was procured. Consequently, ESMD looked for a possible secondary payload to send to the Moon with the LRO spacecraft. A concept was proposed by the NASA–Ames Research Center to crash the expended Centaur upper stage of the Atlas launch vehicle into one of the poles of the Moon and observe the ejecta plume of that impact with a small spacecraft following behind, unlike the previous Lunar Prospector effort in 1999, which attempted to observe the ejecta plume only with Earth-based telescopes. If ice is present on the Moon, it was hoped that we would observe it in this plume. This add-on mission was called the Lunar Crater Observation and Sensing Satellite (LCROSS).[23] It was something of a gamble, since it might miss any putative ice deposits or fail to see those that are present, but was thought to be worth trying. As it turned out, this mission would be the first—and to date, the only—ground truth for the lunar poles that we would get.

The Decline and Fall of the VSE

As the momentum to demote the Moon's role in the Vision grew, Project Constellation started to run into technical issues and mass growth, and consequently, budgetary problems. One issue was the sizing of the new Orion spacecraft. In order to accommodate its larger crew with amenities such as a kitchen and a toilet, the decision was made during the ESAS to adopt a 5-meter diameter for the vehicle (the Apollo command module was 3.9 meters in diameter). This larger spacecraft might have made travel around cislunar space more enjoyable, but such comfort came at a serious cost. The increase in size and mass meant that Orion outgrew its Ares I ("the stick") launch vehicle. Despite an attempt to solve this problem by adding another solid rocket motor segment to the Ares I, now with a five-segment first stage, it was found that to achieve orbit, the vehicle would need to fire the service module engine, much as the shuttle orbiter used its orbital maneuvering engines to finalize its attainment of LEO. This issue was accompanied by concerns over a high-frequency vibration called thrust oscillation during the burn of the Ares I first stage solid-propellant motor; it was feared that this thrust oscillation could temporarily incapacitate the crew during critical abort phases of the ascent. Although these problems all had solutions, the problem was that they did not have any "no-cost, no-mass" solutions.

The basic problem with Orion was that it was oversized for its role as simple transport to and from LEO to support the ISS (part of the ESAS ground rules) and even as a cislunar vehicle. Worse, it was undersized in its role as a Mars spacecraft, being useful for only two phases of the mission: crew departure from the Earth and aerothermal entry upon return. For true, long-duration flights, Constellation would need to carry a separate habitation module. But such a requirement negated the rationale for providing Orion with a kitchen and toilet, which drove its larger size to begin with. Thus, we were (and still are) developing a new human spacecraft that was simultaneously too big for its early uses and too small for its intended later one.

As technical issues grew, the agency's annual budget requests began to increase. When budgetary increases failed to materialize, the scope of agency activities decreased. An early casualty of this new austerity was the lunar robotic program. The second robotic mission to the Moon was to have been a surface lander and rover, designed to follow up on the water discoveries from orbit and measure the type and quantity of water present in the surface, critical information needed to use the resource. Other robotic missions were designed to emplace infrastructure such as communications relays so that landings at the poles and on the far side could be undertaken, and to test resource extraction techniques such as water production on the surface. Because of budgetary pressures caused by Constellation's development problems, all these missions were deferred to "later," which became "never." This deferral of the robotic program was a blatant neglect of the specific direction within the VSE that a "series of robotic missions to the Moon be undertaken" as part of lunar return.

Few observers in Washington ever thought that the VSE would be fully implemented with the minimal new investment that NASA had been promised during the Bush roll out. But it seemed to many that the agency was not trying very hard to maximize the leverage provided by the use of legacy hardware. The Ares vehicles, although based upon an adaptation of shuttle hardware, required so many modifications that it became a completely new development. And given the problems with accommodating an oversized Orion, most didn't even want to think about developing its necessary companion, the behemoth Altair lunar lander, supersized because of its dual role as a self-contained human lander/habitat and an automated cargo lander. There were increasing complaints about Constellation, initially from the space community peanut gallery. Over time, criticisms started showing up in congressional hearings. It didn't help matters that some senior NASA personnel were incapable of explaining exactly why we were going to the Moon in the first place, including some who had been assigned this task as part of their job description.[24]

Meanwhile, progress continued on the two robotic lunar missions that had already been approved. In the fall of 2008, I once again made the long journey to India, only this time to the SHAR complex north of Chennai, on the eastern coast where India launches its rockets. SHAR is located on a flat, marshy coastal plain, similar in setting and ambiance to our own Cape Canaveral. On October 22, 2008, after a few days of constant monsoon rain, we finally launched Chandrayaan to the Moon. As it arced eastward over the Indian Ocean, I was able to catch a quick glimpse of the departing rocket through a miraculous break in the cloud cover.[25] Following a four-day journey, Chandrayaan inserted into orbit around the Moon and began transmitting data back to Earth. I was at the Mission Control Center in Bangalore for our initial data collect in early November, as a single strip showing some lunar craters near the north pole was downloaded. With our instrument working, we began our first mapping cycle in early February 2009 and over the course of the next month, acquired nearly complete maps of both poles. It would take several months of analysis before we could understand what all the data meant—that water ice does exist in quantity in some of the craters near the poles (figure 5.1).[26]

The election of Barack Obama as president in November 2008 led to new uncertainty about the fate of Project Constellation and the VSE. During the election campaign, Obama made ambiguous statements of support for the space program, first suggesting that money expended on space

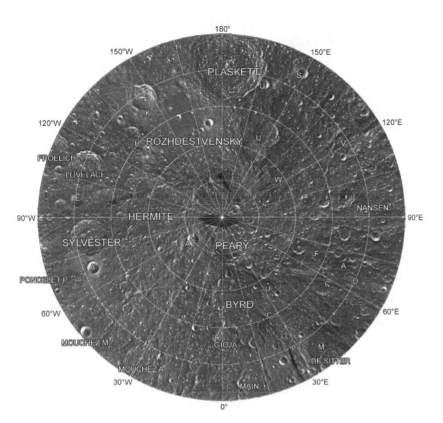

Figure 5.1. Mini-RF radar mosaic of the north pole of the Moon. Small craters with bright interiors near the north pole (arrow) are probably filled with water ice. More than a billion tons of water ice are likely available at each pole.

might be better spent on "education," but rapidly changed his tune during an appearance in the electoral vote-rich, critical state of Florida, where he pledged support for Project Constellation. Space supporters were cautiously optimistic upon his assumption of office. Mike Griffin hoped to stay on as NASA administrator but that was not in the cards and his resignation was accepted. While searching for a permanent replacement, Acting NASA Administrator Chris Scolese testified to Congress that he did not know what "return to the Moon" meant in the context of his agency's activities.[27] At the time, I thought that this statement by the head of the agency assigned the job of implementing the VSE was the absolute nadir of the American space experience but unfortunately, even lower points were to follow. Eventually, former astronaut and Marine general Charles Bolden was named as the new head of the agency, with space advocate and "Astro Mom" Lori Garver assigned as deputy administrator. Completing this cast of characters was Presidential Science Advisor John Holdren, neo-Malthusian environmentalist and critic of human spaceflight.[28]

The first space policy decision of the new administration was to appoint a committee to review the space program and make recommendations on whether to continue current efforts or to reorient its goals and/or the means to implement them. This committee, named for its chairman Norman Augustine, formerly CEO of Lockheed-Martin, should not be confused with the earlier, 1990 Augustine Committee.[29] This new Augustine committee conducted "independent" cost analyses, performed by the Aerospace Corporation, of current NASA programs, with an eye toward possible alternatives. The committee worked throughout the summer of 2009, holding meetings and listening to testimony from agency engineers on progress with various developments underway as part of Project Constellation. The lack of fulfillment of requested levels of funding was a constant refrain. During their meetings, the Augustine Committee also heard testimony from other parts of the agency, including engineers working on Ares alternatives and the value of using in situ resources to make consumables and propellant on the Moon. You will look long and hard in the committee report to find mention of this evidence, which led many of us to suspect that the committee was well on its way to a predetermined conclusion.

The 2009 Augustine committee report,[30] given the grandiose title *Seeking a Human Spaceflight Program Worthy of a Great Nation*, outlined three possible paths forward. One path emphasized a human Mars mission, deemed technically a bridge too far. Another path described a return to the Moon, deemed too old hat. The third alternative outlined what was called the Flexible Path, deemed just right. In contrast to the first two options, Flexible Path advocated journeys beyond LEO to a variety of destinations beyond the Moon but short of the surface of Mars. Such targets included an L-point, a near Earth asteroid, or one of the moons of Mars. You might recall that this was the same "path of progress" advocated by NASA's Decadal Planning Team.[31] The perceived advantage of Flexible Path was that all of its possible destinations are low gravity objects, so that deep space systems could be developed incrementally without the need to simultaneously develop an "expensive" lander spacecraft. The committee had detailed cost estimates for the various options performed by the Aerospace Corporation to buttress its conclusion that no viable and affordable path forward was possible under the budget guidelines given to them by the White House.

The reaction to the work of the committee was mixed. It was widely and incorrectly interpreted as a slapdown of the Constellation architecture. In fact, the report noted that the chosen Constellation architecture would create the capabilities claimed for it. However, costing estimates suggested to the committee that an additional $3 billion per year was needed to meet the chosen schedule goals of Constellation. Attention mainly focused on the Augustine committee's Flexible Path architecture, one that promised technology development in the near term and missions to unspecified destinations sometime in the future. Some thought this was a great approach, while others pointed out that nebulous goals and indefinite timelines are, in general, not a good recipe for a space program "worthy of a great nation."

As always with committee reports, the devil was in the details. Cost estimates provided to the committee by the Aerospace Corporation included excessively large margins and totals came in much higher than other analysts estimated. Moreover, the committee had been presented with evidence showing that modifications to Constellation and other alternatives, such as shuttle side-mount for heavy lift, were possible and affordable without a funding augmentation. Leverage provided by and capabilities created through the use of the resources of the Moon to enable both lunar and martian missions were documented and presented to the committee. Yet, none of these alternative options were given serious consideration.

NASA Administrator Charles Bolden was on record making public comments suggesting he was not enamored of the VSE goals, particularly the one involving the Moon. He was critical of lunar return and indicated that while he was strongly in favor of a human mission to Mars, he believed that it was far away in cost and time. But no matter what new direction human spaceflight took, Bolden stated that he was against any future change to that direction.[32] President Obama's science and technology advisor, John P. Holdren, indicated his desire to make NASA principally responsible for global monitoring of the Earth, with an emphasis on the tracking of climate change from space. It was clear that a correlation of forces was assembling to significantly change the direction and outlook of NASA and the US human space program.

President Obama's April 2010 speech at Kennedy Space Center in Florida outlined his administration's new space policy.[33] At first glance, it appeared to embrace the Flexible Path of the Augustine committee. Obama called for spending on technology development, to be followed by human missions to a near Earth asteroid. He also called for increased efforts to develop commercial capabilities to launch payloads to low Earth orbit. A planned return to the Moon was dismissed with the trite phrase, "we've been there—Buzz has been there," a reference to Buzz Aldrin, who had flown on Air Force One to Florida with the president, apparently giving him the benefit of his vast space expertise during the flight.

The announcement of this new path effectively ended the VSE. More significantly, it was also the end of any strategic direction whatsoever for the American civil space program; that direction had been replaced with rhetoric and flexibility. The promise of spaceflight in the future became the stand-in for real spaceflight in the present. Instead of a mission for people beyond LEO, we were given vague promises of "a spectacular series of space *firsts*." Inconceivably, a relatively small, preexisting program designed to help develop commercial resupply of cargo to and from ISS was heralded as the centerpiece of America's space program—the "new" direction. Gone was the concept of creating a lasting, sustainable spacefaring infrastructure. Back was the template of one-off, stunt missions to plant a flag and leave footprints on some new, exotic, faraway target—it

didn't matter which one—sometime in the distant future, the all too familiar "*exciting* space program."

What many forgot or chose to overlook was that with large bipartisan
majorities, the VSE had been endorsed by the Congress in two separate
NASA authorization bills, once in 2005 and again in 2008.[34] Understand-
ably, Congress did not react favorably to Obama's new direction for the
civil space program. In the new 2010 authorization bill, Congress laid out
some surprisingly detailed specifications for a new heavy lift launch vehicle.
It directed NASA to transform the planned Ares rockets of Constellation
into a new heavy lift launch vehicle to be called the Space Launch System,[35]
or SLS, dubbed the Senate Launch System by its critics. While enthusiasts
for the new direction decried the Congressional actions as pork, the simple
fact was that many on the Hill, sensing that a critical national capability
was being irretrievably lost, were concerned with the unabated, scheduled
retirement of the shuttle. Orion was retained as the program to develop a
new government-designed-and-run human space vehicle.

Interestingly, the resulting 2010 NASA authorization bill kept all the
potential destinations of the old VSE, including the surface of the Moon,
something else that many have ignored. Despite the fact that this bill was a
partial repudiation of his proposed space policy, President Obama signed
it into law. NASA architecture teams examined possible human missions
beyond LEO, including to an L-point and near Earth asteroids, but an
achievable mission that would materially advance our spacefaring capability could not be identified. To disguise the embarrassment of not finding
an asteroid that a human crew could reach, the agency embraced the pre-
posterous idea of capturing a small asteroid and returning it to an orbit
around the Moon: the Asteroid Return Mission, or ARM.[36] At that point, the
space rock would be accessible to a human crew using the Orion spacecraft,
launched on the new SLS vehicle. This concept was roundly criticized, and
most space stakeholders reviled and rejected it, except for those who had
advanced the idea in the first place. Congress has yet to embrace the ARM
and is split on possible future destinations, although it is still considering a

possible Mars flyby, a Phobos landing, an L-point mission, or even (gasp!) lunar return. Everything is up in the air—and we are going nowhere.

Regrouping

So we arrive at the present: a space program without strategic direction and an uncertain future. We have seen the confusion and chaos that resulted from two different presidential attempts to set a long-term direction for the space program. These efforts were torpedoed by a variety of effects and events. Primarily, it was a lack of understanding of the objectives of lunar return or disagreement with them. We had experience with humans on the Moon during the Apollo program, and many, including some inside the space program, could not imagine anything that people could do there that was different than what the Apollo astronauts did—hop around, collect some rocks, ride in an electric golf cart, and fall down a lot. The characterization of Project Constellation as "Apollo on steroids" did nothing to convince and educate people that there were new and exciting possibilities involved in lunar return. Those who were offended by the idea that we were simply "repeating the Apollo experience" on the Moon did not notice that in their alternative program, they were endeavoring to perform that very experience on Mars, with a similar flags-and-footprints extravaganza.

In the years since the demise of Constellation, a common complaint is that President Bush and the Congress did not adequately fund the VSE. This is untrue. On the rollout of the VSE, the amount of funding that NASA was to receive was specified: an additional $1 billion, spread out over the next five years (2005–2009), after which the agency budget was to rise only with inflation. NASA received this funding, although not in equal amounts over that period of time (Figure 5.2).[37] The additional funding needed to develop the new CEV and launch vehicles was to come from the "wedge" produced as a result of money freed up by the shuttle retirement and the rampdown of the ISS program. Additionally, Congress formally endorsed the goals of the VSE in two different authorization bills; those two bills passed with large majorities by a Congress under the respective

NASA budget FY 2004-2009

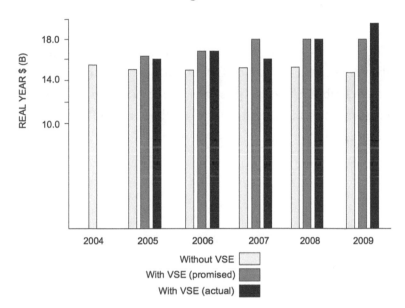

Figure 5.2. Funding for NASA during the first five years of the Vision for Space Exploration. The light color is the agency funding without the VSE, the middle color is promised funding (an additional $1 billion, spread over five years with allowance for inflation) and the dark color is actual funding. In contrast to prevailing myth, NASA received all of the VSE funding that it was promised.

control of both Republicans (2005) and Democrats (2008). Thus, the VSE was a presidential proposal, adopted on a bipartisan basis as national policy by Congress and funded at the levels promised. NASA was tasked with coming up with a plan to return to the Moon under those boundary conditions, not to devise an unaffordable architecture and then whine about not having enough money to do it.

As we have seen, new data for the poles of the Moon show that the critical resources of energy and materials are available there in usable form. This appreciation requires that we rethink our purposes in space and on the Moon. The VSE was an attempt to test a new paradigm of space operations—instead of bringing everything we need with us from Earth, we would learn how to access and use what we find in space to provision ourselves and to create new capabilities there. As we endeavor to break the logistical chains of Earth and become a true spacefaring species, this effort holds the potential to give us unlimited capabilities in space.

What is the best path forward? Was the original plan to use the resources of the Moon to create new spaceflight capability the right idea? What can we do to advance our "reach" beyond LEO into the solar system? Why is such a thing even desirable? These are questions I hope to answer in the next few chapters as I examine the facts, the potential, the hype, and the possibilities for the future of the American civil space program.

6

Why? Three Reasons the Moon Is Important

Throughout all of the various attempts to give our national space program a long-term, strategic direction, the Moon has waxed and waned in significance. Despite many attempts over the last thirty years to ignore it or focus exclusively on robotic space science or human missions to Mars or the asteroids, the logic of lunar return has not been refuted. Undeniably, the Moon will figure prominently in any plans for human spaceflight beyond low Earth orbit, if not by the United States, then by some other nation with the foresight and the will to take the lead.

Previous attempts to define *the* "mission" on the Moon—the quest for various rationales for lunar presence—has produced multiple themes, goals, and objectives, the most infamous being the six themes and 186 objectives adumbrated at the 2006 NASA Exploration Workshop.[1] It really isn't that complicated. I will attempt to cut through this programmatic fog in order to examine the fundamental reasons why the Moon is not merely important but also critical for the development of permanent spaceflight capability. Whatever long-term space goal we adopt, the Moon will play a key role in enabling us to achieve those objectives. The value of the Moon lies in three principal attributes: It's close, it's interesting, and it's useful. I will examine each attribute in turn, evaluating its significance to the development and exploration of space.

It's Close: The Value of the Moon's Proximity

Unlike most other space destinations, the Moon is Earth's companion in space. The Earth-Moon system orbits the Sun as a single planet. Thus, the Moon is always accessible from the Earth. This is in marked contrast to other deep space targets such as planets and asteroids, all of which have independent solar orbits and thus, are optimally accessible only during certain short, periods called "launch windows." In the case of Mars, good launch windows, those requiring the minimal amount of energy for transfer, expressed as "delta-v" or change in velocity, occur about every twenty-six months. Other targets, such as near Earth asteroids, may have more frequent windows separated by months but lasting only a few hours to a few days; some have even fewer launch windows.

The Moon is always available. Fifty years ago, the Apollo launches were scheduled within very tight launch windows because the Lunar Module had to land on the Moon in the early morning hours, when cast shadows make the surface relief stand out clearly. In the case of a future lunar outpost, one of the first items to emplace on the surface will be a beacon, a radio device that allows future landers to land completely "blind" at any time of the lunar day or night. Departures and arrivals will be conducted for convenience, with timing imposed not by celestial mechanics but by the operational schedules of the flight systems manager. A series of radio beacons would enable the development of a completely automated flight system, one that could transport goods and people between Earth and a lunar outpost.

The Moon is accessible via many different orbital approaches (figure 6.1). Direct paths, requiring the maximum amount of velocity change (delta-v), are possible from Earth, resulting in transfer times on the order of three days; minimal modification permits lower total energy requirements and adds another day or so to transit time. Staged approaches can be conducted using the L-points or low lunar orbit as a staging location. The advantage of such an approach is that assets and pieces of a complex system can be assembled at a staging node, with the surface mission conducted from that point. The Apollo system used low lunar orbit (100 km circular) as a staging area. Staging from one of the L-points—usually L-1, about

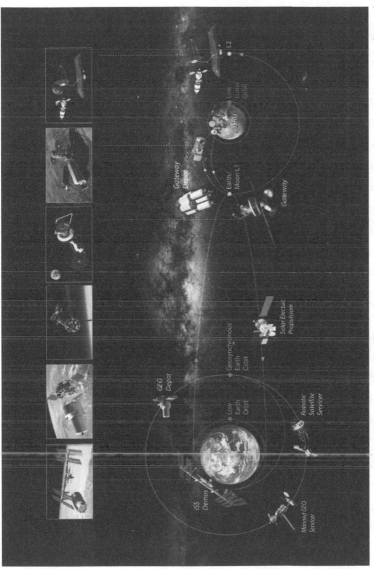

Figure 6.1. Zones of cislunar space. Low Earth orbit (LEO) is the location of the International Space Station and the limit of most human missions. Geosynchronous orbit (GEO) is the location of communications and weather satellites. The Earth-Moon L-1 and L-2 points are possible staging locales for trips to and from the Moon. Low lunar orbit and the surface are within the gravity well of the Moon.

60,000 kilometers above the center of the near side of the Moon—has many benefits, including its utility as a marshaling area for lunar exports when water production meets that level and for constant line-of-sight communications with both Earth and Moon. Finally, it is possible to send large payloads of cargo via "slow boat" transfer routes using efficient, low-thrust, high-energy techniques, such as solar electric propulsion. These transfers spiral out to lunar distances over periods of weeks to months. But while they pass through the Van Allen radiation belts multiple times, they impose no hazards because they carry only cargo and not people.

If problems arise during lunar journeys, the return to Earth takes only a few days. On planetary missions, a return to Earth may take many weeks to months, if possible at all. An abort capability is critical for the planning of human missions. This was demonstrated most dramatically during the flight of Apollo 13 in April 1970.[2] An explosion of an oxygen tank in the Service Module crippled the spacecraft's electrical system, making the Command Module inoperative. Because the Lunar Module was still attached to the vehicle (they were on their way to the Moon), the crew was able to use it as a "lifeboat" to survive for the three days it took to swing around the Moon and return to Earth. The ability to abort a flight in progress, for safety or other operational reasons, distinguishes the Moon from other planetary destinations. This benefit is an enormous advantage during the early stages of space development, as the reliability of new systems has yet to be demonstrated. Catastrophic loss of crew can bring a nascent program to a halt, and in some cases, result in its early termination.

Ease of communication with the Earth is another advantage of the Moon's proximity. The leverage provided by this short time-delay ranges from the merely convenient to the operationally essential. For typical human operations on the Moon or at lunar distances, round-trip radio time is a bit under three seconds, a noticeable but easily handled delay, as listening to any of the Apollo audio files will attest. The critical value of the short time-delay of lunar distance comes with robotic teleoperation. As will be discussed in more detail later, a strategy for acquiring early operational capability on the Moon will come from the emplacement and use of robotic assets. These robots will prepare and construct the infrastructure of the lunar outpost, as well as begin work on resource harvesting, water extraction, storage, and processing, work that can be operated, or at least supervised, remotely from Earth. Because of the tens-of-minutes time delay for radio propagation, the remote operation of machines on Mars makes it difficult to accomplish even the simplest of tasks. In contrast, the proximity of the Moon permits us to operate assets on the lunar surface in near real time.

The many advantages of the Moon's closeness make it a logical and useful destination in any trans-LEO human spaceflight architecture. The creation of capability in space will be accomplished more easily and safely by first learning how to operate in space at lunar distances. With experience and competence acquired on the Moon, we will be more confident and skilled when we move outward to more distant destinations. By using the Moon to learn these skills and techniques, we learn how to crawl before we attempt to walk.

It's Interesting: The Scientific Value of the Moon

The Moon offers scientific value that is unique within the family of objects in the solar system.[3] It is a recorder of history and process, an ancient world containing materials unprocessed since their formation more than four billion years ago. The Moon records its own history and the history of the universe around it. Its environment permits unique experiments in the physical and biological sciences. Additionally, it is a natural laboratory for understanding the processes that created our solar system and that currently drive the geological evolution of the planets.

The Moon has undergone a complex and protracted geological history that we can study to understand early planetary evolution. From Apollo data, we found that the Moon is a differentiated object, with a metallic core, mantle, and crust. Its segregation into this tripartite condition was the result of global melting early in solar system history. If a body as small as the Moon could undergo global differentiation, it is likely that all the terrestrial planets did likewise. The study of early lunar geologic history is a guide to the interpretation of the history of all the rocky planets, and the Moon records events of an epoch for which evidence has been erased from the eroded, dynamic surface of the Earth. After this differentiation, the Moon underwent a protracted impact bombardment, hit by objects from the microscopic to the asteroidal; these collisions formed craters that span similar size ranges. While we understand the impact process in broad outline, details of the physical and compositional processes remain obscure, especially questions about how they scale with size. The Moon's abundant craters (figure 6.2), on display for our study and enlightenment, offer innumerable examples of this process.

Billions of years ago, internal melting of the mantle of the Moon produced copious iron-rich magmas that rose upward to the surface and erupted as vast sheets of basaltic lava. These lavas make up the lunar maria, the dark smooth lowlands of the Moon. They are concentrated on the near side (for reasons that still elude us) and are made up of hundreds of individual flows with differing compositions, volumes and ages. By understanding the sequence of lavas over time, their source regions, and changes in composition, we can reconstruct the thermal and compositional evolution of the lunar deep interior. Again, because volcanism is ubiquitous on the terrestrial planets, knowledge of the lunar experience helps us to better understand this process across the solar system.

The principal geological process on the Moon for the last three billion years is bombardment by a constant micrometeorite "rain" of tiny particles. The flux of debris acts as a giant "sandblaster," grinding surface rocks into a fine powder. This layer of disaggregated rocky debris, the regolith, is exposed to space and thus, implanted with particles from sources external to the Moon. Because the Moon has no atmosphere or global magnetic field, plasmas and streams of energetic particles from the Sun, and the universe around us, impinge directly on its surface, becoming embedded onto these lunar dust grains. Thus, the Moon contains a unique, detailed record of the output of the Sun and galaxy through geological time.

The solar wind is the most common source of particles, a stream consisting mostly of protons that collide with and stick to the lunar dust grains. As this process is constant, particles from the Sun emitted at varying times in history may be recovered from the ancient regolith and used to reconstruct the output of the Sun and galaxy as it was in the distant geological past. A special case occurs when an ancient regolith is buried by a lava flow. In this instance, the covered regolith becomes a closed-system, shut off from

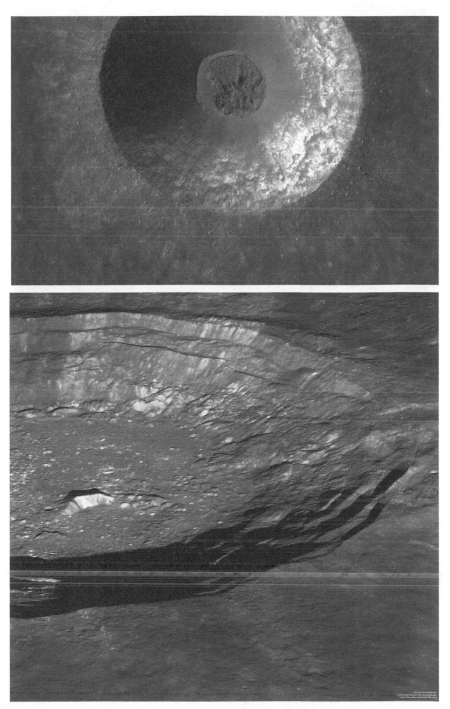

Figure 6.2. Examples of fresh lunar craters. Rümker E (38.6°N, 302.9°E; 7 km diameter) is a simple crater, with a bowl shape and small, flat floor. Large blocks are visible near its rim crest. The complex crater Aristarchus (23.7°N, 312.5°E; 40 km diameter) shows wall terraces (from slumping after crater excavation), an extensive flat floor (impact melt sheet) and a central peak (brought up from the deep crust).

further particle implantation. The solar wind gases, preserved in such a closed-system, record a "snapshot" of the ancient Sun, dated by the ages of the bounding rock units above and below the ancient paleoregolith.

Accessible regoliths on the Moon cover a time range of at least the last four billion years. The Sun is the principal driver of Earth's climate, and by recovering solar output over time, a record unavailable anywhere on Earth, we can understand its cycles and singular events for the duration of the history of the solar system. Some initial results, from our study of the Apollo samples, suggest that the ancient Sun had a different composition of its nitrogen isotopes than it does now, a puzzling result not predicted by existing theories of stellar evolution. What other new and unexpected secrets of the Sun and stars lie embedded on the Moon, awaiting discovery?

Because of the antiquity of the Moon, and its proximity to the Earth, the lunar surface retains a record of the impact bombardment history of both bodies. We know that the collision of large bodies has had drastic effects on the geological and biological evolution of the Earth and occur at quasi-regular intervals.[4] Because our very survival depends on our understanding the nature and history of these collisions as a basis for the prediction of future events, the impact record on the lunar surface is critical to our understanding of this hazard. By dating a large population of individual craters on a surface of known age, we can establish whether the periodicity of the impact flux is real. Such periodic impacts may have driven the process of evolution on Earth. These studies could uncover fundamental, unknown aspects of the history of life on Earth and in the solar system.

With no ionosphere, and a far side that is the only known area in the solar system permanently blocking the radio noise and static of Earth, a radio telescope on the far side of the Moon can examine low frequency wavelengths that are impossible to detect from Earth's surface or in LEO. The seismically quiet lunar surface permits the construction of extremely sensitive and delicate instruments, such as interferometers at optical wavelengths. An array of such telescopes could achieve resolutions at the micro-arc second level, allowing the direct observation of phenomena such as star spots and the hemispheres of terrestrial planets in nearby systems. Such capabilities would revolutionize our understanding of the evolutionary paths of stellar and planetary systems.

Finally, the environment of the Moon is itself a scientific asset of great value. The hard vacuum and extreme thermal regime permit unique material science experiments. The low gravity of the Moon allows us to quantify the effects of fractional gravity on physical and biological phenomena. The Moon is an isolated and sterilizing environment, permitting experimentation with hazardous materials and processes. Facilities on the lunar surface allow us to conduct dangerous or hazardous experiments that would be unwise to pursue on the Earth. These unique properties make the Moon an unparalleled asset for scientific experimentation and laboratory work.

It's Useful: The Utility of the Moon

While the previous two attributes of the Moon are extremely important, its greatest value is its capacity to create new spacefaring capability through the exploitation of its material and energy resources. The idea of using the materials of other worlds to provision ourselves, and to supply and support spaceflight, is a very old one, but to date, it has not been attempted. Yet, development of this single activity could completely change the paradigm of spaceflight. Currently, anything that we need in space must be transported to Earth orbit at enormous cost, usually on the order of at least $1,000–10,000 per kilogram. This high cost applies to everything: It costs the same amount of money to launch a kilogram of high-technology electronics as it does a kilogram of water. If we could provide low information density materials (like water, air, and rocket propellant) from local sources already present in space, we could accomplish *much more* for less money. In a nutshell, this is the driving motivation for the use of off-planet resources, or, in the term used in the business, in situ resource utilization (ISRU).[5] This is a skill that we must master in order to become a truly spacefaring species.

Although the physics and chemistry of extracting and using the resources of the Moon are simple and straightforward, there has been great resistance to incorporating ISRU into any spaceflight architecture. There are many reasons for this attitude, ranging from unfamiliarity with the processes involved to a natural and at least partly understandable conservatism in engineering design. For initial ISRU efforts, we would only undertake the simplest processes, such as bulldozing regolith to make blast berms around landing pads and to cover habitats for radiation shielding, along with heating polar regolith to extract water ice. These are minimal, low-risk activities that provide useful products and pieces of outpost infrastructure. The techniques needed to begin ISRU are no more complex than everyday eighteenth-century industrial processes.

The resources of the Moon are simple and require minimal processing. First, bulk regolith (soil) has many uses as thermal and radiation shielding and for construction. Although loose soil can be used as is, regolith can also be fused by microwave sintering or passive solar thermal heating (such as a concentrating mirror) into ceramics or aggregate for building material. Roads and landing pads can be manufactured by sintering the regolith in place using a microwave-heating element mounted on a rover.[6] Microwaves fuse loose regolith into brick and ceramic because of the fine-scale, vapor-deposited free iron that coats the surfaces of dust grains. This coating permits RF energy to be efficiently coupled and transferred into heat, so that the grain boundaries fuse together to make glass. A microwave with a power level comparable to a kitchen oven can fuse the upper surface into a paved road or landing pad several centimeters deep. Fused regolith structures can be made as large or as long as needed. Structures and pieces can be produced with 3-D printer technology using fine regolith as feedstock.

The Moon's poles possess critical resources needed for long-term human presence on the Moon and in space. They have two key attributes that the rest of the Moon does not possess: water ice (and other volatile substances) and areas of near-permanent sunlight. We have verified the presence of water ice using several techniques of remote sensing, including hydrogen detection, near-infrared and ultraviolet reflectance, laser albedo, radar, and a physical impactor. In addition to water—the most cosmically abundant volatile substance in the solar system—other volatile species are present in the polar ice, including methane (CH_4), carbon monoxide (CO), ammonia (NH_3), hydrogen sulfide (H_2S), and some simple organic molecules. All of these volatile substances can be chemically processed to help support a human presence on the Moon.

Questions remain over how much water and other volatiles are present in total, on their distribution laterally and vertically, and over what physical form the different chemical ices take. These volatiles probably come from sources external to the Moon—the impact of water-bearing objects, such as cometary nuclei and volatile rich meteorites. As such, they are deposited in extremely small amounts, in a vacuum and over a very long period. The likely nature of such a deposit would be a very porous mixture of dust grains and amorphous (noncrystalline) ice. In astrophysics, such a compositional fabric is called a fairy-castle structure and is a common state of materials in space.

The dark areas where ice is stable are extremely cold, always less than −169°C (104 K), but in some cases as cold as −248°C (25 K) and widespread at both poles. These dark areas are typically found in crater interiors but in some cases as extended regions of shadow. The "cold traps" are all equally likely to contain ice, but current evidence suggests, for reasons we do not fully understand, that the ice is distributed heterogeneously (see figure 5.1). In addition, because lunar soil is an excellent thermal insulator, it is possible that extensive deposits of ice might be present in the shallow subsurface, in areas that receive partial solar illumination.

We need to survey the potential mining areas to determine their content and grade. This is best accomplished by using a small robotic rover that traverses the polar areas and measures ice content and composition over many locations. The dark areas are close to the lit regions, as the grazing sunlight at the poles, both illuminates and shades. Although there are no areas of "permanent" sunlight, certain regions near both poles have been found to be in sunlight for more than 90 percent of the lunar year.[7] Solar arrays mounted on a high mast could be in sunlight for longer periods; this possibility is a subject for current research. An outpost located in these areas would be able to generate electrical power on a nearly constant basis, with periods of darkness bridged by power storage, such as the use of a rechargeable fuel cell.

Another advantage of these "quasi-permanent" sunlit areas is that they are thermally benign. At the equator of the Moon, the surface is heated during the daytime, which is fourteen Earth days long, reaching temperatures of up to 100°C. During the coldest part of the nighttime, also fourteen Earth days long, the surface may assume temperatures as low as −150°C, a 250° swing from the hottest part of the day. The high temperatures of lunar noon put stress on systems designed to keep machinery cool, while the cold night temperatures require moving parts to be heated. Within the sunlit areas near the poles, illumination is always at grazing incidence—that is, the Sun circles around near the horizon—and maintains the surface temperature at a near-constant −50°C. In such an environment, minimal power is required to maintain thermal equilibrium for complex machinery. Along with the pervasive presence of highly abrasive dust that can wear down parts and make machinery inoperative, the extreme thermal environment is one of our biggest technical challenges in developing the resources of the lunar poles. Mitigating strategies for each of these difficulties are currently the subject of intensive research.

In addition to the constant solar power available at the poles, the Moon contains substances that, in the future, may be used to generate energy for use on the lunar surface and in space. Several regions of the western near side contain elevated amounts of the radioactive element thorium, which can be used to fuel nuclear reactors to generate electrical power. Via several nuclear reactions, thorium breeder reactors can produce their own fuel, making it possible for us to construct space reactors on the Moon. The use of nuclear power would allow us to survive the long lunar night and permit habitation of equatorial and mid-latitude regions of the Moon. The availability of abundant power also enables large-scale industrialization of the Moon.

In the more distant future, some have proposed that the rare isotope helium-3, implanted in the lunar regolith by the solar wind, could be harvested to generate electrical power in a relatively "clean" nuclear reaction, one that does not generate excess neutrons and "dirty" reaction products.[8] The fusion of deuterium (hydrogen-2) with helium-3 produces fewer neutrons and positively charged He ions, permitting the efficient conversion to electrical power over the standard deuterium-tritium (^{2}H-^{3}H) fusion. In fact, a variant of this process, whereby helium-3 fuses with itself (^{3}He-^{3}He), produces no harmful by-products at all. Potentially, helium-3 fusion could solve the world's energy problems if a suitably large source of the isotope could be found   it is present on Earth as a component of natural gas, but in extremely small amounts.

It has been proposed that we mine the lunar regolith for helium-3 and import the product back to Earth for commercial electrical power generation. The difficulty with this idea is twofold. First, we do not yet have reactors that can burn helium-3 nuclear fusion fuel. It takes a great deal of energy to start this reaction and then to contain and control it; no fusion reaction to date has achieved "breakeven," the point at which the fusion reaction liberates more energy than it takes to start it. Research on this problem has been going on for decades; it is unlikely that we will see commercial applications of fusion power generation for many years. Second, although there is helium-3 in the lunar regolith, it is present in concentrations of less than about twenty parts per *billion*. This low concentration is for sampled sites in the lunar equatorial maria; we do not yet know the concentration of helium-3 in the polar volatiles. Extracting helium-3 from the mare regolith will require the mining and processing of hundreds of millions of tons of regolith, a scale of resource processing that may eventually occur, but certainly not in the early stages of lunar habitation. The mining of helium-3, often alluded to as the ultimate "pay dirt" on the Moon, is not likely near-term (~20 years) but may turn out to be significant in the multidecadal time scales of future lunar development.

Water is the most useful material in space. In its native form, we can drink it and use it to reconstitute food, cool equipment, and jacket habitats for radiation protection, as well as for hygiene and sanitation needs. An electrical current can disassociate water into its component hydrogen and oxygen. These gases can be stored and used; oxygen can be used for breathing, and both gases can be recombined in a fuel cell to generate electricity. Used this way, water is a medium of energy storage. Finally, the hydrogen and oxygen can be cooled into cryogenic liquids and used as rocket fuel, the most powerful chemical propellant known. Because of its utilitarian value, water is truly the "currency" of spaceflight.

The real lunar "El Dorado" consists of the water ice and the permanent sunlight near the poles. It is a location known to contain resources of material and energy that we can access and use. It is a place where we can learn the skills and technologies needed to become permanent residents of space.

Why Not Mars?

Virtually the entire space community, from those inside the agency to others working on spacecraft, missions or data analysis, presume that Mars is the "ultimate goal" for human spaceflight.[9] In 1965, the imaginative pull that decades of science fiction and speculation about Mars as an Earthlike planet had dealt us were dashed when we found by direct investigation that the real Mars is a distant, cold, dry desert, with virtually no atmosphere. Subsequent missions over the years have shown that it may have been warmer and wetter in the past, which led to the idea that microbial life might have originated there. This single idea is largely responsible for the subsequent fixation on Mars as the "next destination" for humans in space. The obsession with "searching for life elsewhere" has hijacked our thinking about the future of people in space. It is virtually impossible to advance an idea or concept involving people at some space destination other than Mars, without proving that it "feeds forward" to our "ultimate destination."

We do not know how to send people to Mars at this time. The difficulties with a human mission to Mars fall into several categories: technical, programmatic, and fiscal. The manned space program has conducted long-duration spaceflight, built a heavy-lift launch vehicle, and conducted landings on the Moon. But for a variety of reasons, getting humans to Mars is much more difficult. Mars is much farther away from Earth, varying between 140 to 1,000 times (55 to 400 million km) the distance of Earth to the Moon (400,000 km). No known trajectory can shorten the months of transit; most robotic missions take nine months. Although issues of crew deconditioning caused by microgravity appear to be mostly resolved from flight experience on the ISS, months of exposure to hard cosmic radiation and the occasional possible solar particle event, requires some type of shielding. People need to breathe, eat, and drink, so those consumables must be carried with them. Mars is bigger than the Moon (its gravity is about 3/8 that of the Earth, compared to the 1/6 g of the Moon); thus, it requires more energy to descend and land on the surface of Mars. This applies to the return trip as well. The larger gravity well of Mars means that bigger landers and more fuel are needed. Although there is an atmosphere on Mars, it is more than one hundred times thinner than Earth's, so we cannot rely solely on aerothermal entry to slow down the spacecraft; a significant propulsive maneuver is required. This issue, the EDL (entry, descent, landing) problem,[10] is one for which we have no solution at present.

The composition of the martian atmosphere is virtually pure carbon dioxide (CO_2) and thus, not breathable; the thinness of the atmosphere requires people to wear pressure suits. The surface is not completely shielded from cosmic rays and solar UV radiation; Mars does not have a magnetosphere like the Earth, which means that it is a hard radiation environment, limiting the permissible time for surface exploration. The soil on Mars is very fine, probably consisting of clay minerals, and owing to the presence of perchlorates and other highly oxidizing substances, it appears to be highly reactive chemically. If inhaled—and some dust inevitably will be brought into the crew cabin—dust could result in caustic chemical reactions in the bronchia of the crew's lungs. In addition, no one knows how well humans will cope with the reduced gravity of the martian surface after spending multiple months in microgravity.

The biggest problem with a human Mars mission comes right at the beginning. In order to carry the fuel, consumables, equipment, and vehicles that the crew will need for this trip, it will require the launch of between 500 and 1,000 metric tons into low Earth orbit.[11] Getting this mass into orbit will require eight to twelve launches of a heavy lift rocket, each carrying about

130 metric tons to space. The vast bulk of this mass is the fuel required for the trip, most of which is burned at the beginning of the voyage to insert the vehicle into its planetary trajectory. But that's not the end of the issue. The fuel will probably be cryogenic hydrogen and oxygen; if we use storable propellant, multiply the mass needed by factors of two to three. After it is delivered to orbit to await the eventual mission to Mars, the super cold cryogenic fuel will quickly boil off from solar heating. It would be a race against the clock to get enough fuel in one place in orbit at the right time. For now, we have no solution to the problem of fuel boiling off in the landers, both for descent and ascent, during the cruise phase of a manned mission to Mars.

Other problems arise in terms of the scheduling of the multiple HLV launches and coordinating their payload manifests. Only two HLV launch pads (Launch Complex 39) exist at Cape Canaveral. One is currently unavailable, leased to a private company. Thus, we would need to launch all of these vehicles from a single pad. To get the pieces of the Mars mission in one place and ready to go, we must deal with an enormous scheduling and manifest problem, as well as the logistics of multiple HLV deliveries. After that, the next hurdle would be assembling and fueling the Mars vehicle in space.

The cost of a Mars mission conducted in this manner is estimated at several tens of billions of dollars per trip. Is such a cost politically viable? Regardless of the propaganda spun by a hopeful New Space community, there are no magic bullets to lower this enormous cost. We still need the same mass in LEO, and the "lowering" of launch costs, which in any event is only on the order of factors of two or three at best, might turn a $500 billion mission into a $450 billion mission. For context, we currently spend about $18 billion per year on our civil space program, of which about $8 billion is designated for human spaceflight.

Faced with these realities, it should be evident that Mars is very far from Earth, technically and fiscally. But the hardwired dreams of living on Mars have left space advocates of all persuasions chasing their tails, locked in a 50-year exercise by the promises of politicians or administrators who tell us, "Yes, we will be embarking on a new program to send humans to Mars." What follows, as night follows day, is that people get spun up and start conducting feasibility studies; new vehicles are designed, and lovely color artwork showing people rappelling down the walls of one of the canyons of Valles Marineris is produced. And then, yet again, the dismal mathematics of a Mars mission becomes evident. But, we are told, not to worry: The mission is at least a couple of decades into the future. Somehow, the money and the political support for more money still will magically appear at the right time. Certainly, if we can assemble a Mars advocacy group, one that shows we have clout and that strikes fear in the hearts of our elected officials, we will get more money. To date, these methods and declarations have accomplished nothing. But, our leaders tell us, this will change as soon as we find a way to get the public excited about space—that "excitement" causes money to flow into the space program. After 50 years, is it not time to admit that this approach isn't working?

An article of faith among the true believers is that interest in the Moon and planning for lunar bases has kept them from achieving their lifelong dream of strolling across the red plains of Mars. The reality is exactly the reverse: It is the fixation with sending people to Mars that has kept us from doing any human missions beyond LEO. Looking over the history of post-Apollo planning, from Nixon's Space Task Group in 1969 to the Vision for Space Exploration in 2004, all efforts to get people into trans-LEO space have run aground on the realities of the enormous technical and cost difficulties of human Mars missions.[12] During the VSE, NASA was more concerned with devising a lunar "exit strategy" than it was with getting people back to the Moon in the first place.[13] The dirty little secret is that most politicians love human Mars missions not because they have any desire or interest in *doing* them but because it is an excellent and proven way to keep the space community pacified by selecting a goal that is so far into the future that no one will be held accountable for its continuing non-achievement. What a remarkable accomplishment for America's efforts in space: once we had a real space program that some thought was faked, and now we have a fake space program that many believe is real.

The only way we will ever get people to Mars is through the construction of a transportation system that enables the routine movement of cargo and people throughout space. An Apollo-style crash program to send humans to Mars is highly unlikely to ever materialize. We need to acquire and learn certain spacefaring skills and technologies, including reusable space-based vehicles, staging nodes in deep space, in situ resource utilization, and the manufacture of propellant from water. If we possessed these capabilities, a human mission to Mars, while still challenging, would become more feasible. We can learn those skills and acquire those technologies on the Moon.

Why not Mars? Because it's too far, too difficult, and too expensive.

Why Not Asteroids?

At first glance, it might seem that asteroids, specifically the near-Earth objects (NEO), answer the requirements for future human destinations. NEOs are beyond low Earth orbit, they require long transit times and so simulate the duration of future Mars missions, and we have never visited one with people. However, detailed consideration indicates that NEOs are not the best choice as our next destination in space.

Most asteroids reside not near the Earth but in the asteroid belt, a zone between the orbits of Mars and Jupiter. The very strong gravity field of Jupiter will sometimes perturb the orbits of these rocky bodies and hurl them into the inner solar system, where they usually hit the Sun or one of the inner planets. Between those two events, they orbit the Sun, sometimes coming close to the Earth. NEOs can be any of a variety of different types of asteroids, but are usually small, on the order of tens of meters to a few kilometers in size. As such, they do not have significant gravity fields of their own, so missions to them do not "land" on an alien world, but rather rendezvous and station-keep with them in deep space.

The moniker "near Earth" is a relative descriptor. These objects orbit the Sun just as the Earth does, and depending upon the time of year, vary in distance to the Earth from a few million kilometers to hundreds of millions of kilometers. Getting to one NEO has nothing to do with getting to another, so visiting multiple NEOs during one trip is both difficult and unlikely. Because the distance to a NEO varies widely, we cannot just go to one whenever we choose: Launch windows open at certain times of the year, and because the NEO is in its own orbit, these windows occur infrequently and are of very short duration, usually a few days. Moreover, due to the distances between Earth and the NEO, radio communications will not be instantaneous, with varying time lags of tens of seconds to several minutes between transmission and reception.

Although there are several thousand NEOs, few of them are potential destinations for human missions. This is a consequence of two factors. Because space is very big, even several thousand rocks spread out over several billion cubic kilometers of empty space results in a very low density of objects. Second, many of these objects are unreachable, requiring too much velocity change from an Earth departure stage; this can be the result of either too high of an orbital inclination (out of the plane of the Earth's orbit) or an orbit that is too eccentric (to varying degrees, all orbits are elliptical). These factors result in reducing the field of possible destinations from thousands to a dozen or so, at best.

There are few asteroid targets and it takes months to reach one. Long transit time is sold as a benefit by advocates of asteroid missions: Because a trip to Mars will take months, a NEO mission will allow us to test out the systems for Mars missions. But such systems do not yet exist. On a human mission to a NEO, the crew is beyond help from Earth, except for radioed instructions and sympathy. A human NEO mission will have to be self-sufficient to a degree not present on existing spacecraft. Crew exposure to the radiation environment of interplanetary space is another consequence of long flight times. This hazard comes in two varieties: solar flares and galactic cosmic rays. Solar flares are massive eruptions of high-energy particles from the Sun, occurring at irregular, unpredictable intervals. We must carry some type of high-mass shielding to protect the crew from this deadly radiation, and this "storm shelter" must be carried wherever we go. Because Apollo missions were only a few days long, the crew simply accepted the risk of possible death from a solar flare. Cosmic rays are much less intense, but constant. The normal ones are relatively harmless, but high-energy versions (heavy nuclei expelled from ancient supernovae) can cause serious tissue damage. Although the crew can be partly shielded from this hazard, they are never totally protected from it.

When the crew finally arrives at their destination, more difficulties await. Many NEOs spin very rapidly, with rotation periods on the order of a few hours at most. This means that the object is approachable only near its polar area. Because these rocks are irregularly shaped, rotation is not the smooth, regular spin of a planet, but is more like that of a wobbling toy top. If material is disturbed on the surface, the rapid spin of the asteroid will launch this debris into space, creating a possible collision hazard to the human vehicle and crew. The lack of gravity means that "walking" on the surface of the asteroid is not possible; crew will "float" above the surface of the object, and just as occurs in Earth orbit, each touch of the asteroid surface (action) will result in a propulsive maneuver away from the surface (reaction).

We would need to work quickly at the asteroid because we would not have much time there; loiter times near the asteroid for most opportunities are a few days. Why so short? Because the crew wants to come home. The NEO and Earth continue to orbit the Sun, and we need to make sure that the Earth is in the right place when we arrive back at its orbital position. In effect, we will spend months traveling there in a vehicle with the habitable volume of a large walk-in closet, have a short time at the destination, and then spend months on the trip home.

In general terms, we already know what asteroids are made of, how they are put together, and what processes operate upon their surfaces. Most NEOs will be ordinary chondrites. We know this because ordinary chondrites make up about 85 percent of all observed meteorite falls. This class of meteorite is remarkable not for its diversity, but for its uniformity. Chondrites are used as a chemical standard in the analysis of planetary rocks and soils to measure the amounts of differentiation or chemical change during geological processing. One chondrite is pretty much like all the others.

Questions that could be addressed by human visitors to asteroids concern their internal makeup and structure. Some appear to be rubble piles, while others are nearly solid. Why such different fates in different asteroids? By using active seismometry (acoustic sounding), a human crew could lay out instruments and sensors to decipher the density profile of an asteroid.

Understanding the internal structure of an asteroid is important for learning the internal strength of such objects; this is an important factor in devising strategies to divert a NEO away from a collision course with Earth.

An alleged benefit of travel to an asteroid is that they have resource potential. I agree, putting the accent on the word "potential." Our best guide to the nature of these resources comes from the study of meteorites—NEOs that have already collided with the Earth. The resource potential of asteroids lies not in the chondrites, but in the minority of asteroids that have more exotic compositions. Metal asteroids make up about 7 percent of the population and are composed of nearly pure iron-nickel metal, with some inclusions of rocklike material as a minor component. Other siderophile (iron-loving) elements, including platinum and gold, make up trace portions of these bodies. A metal asteroid is an extremely high-grade ore deposit, potentially worth billions of dollars, if we were able to get these metals back to Earth.

However, from the spaceflight perspective, water has the most value. A relatively rare asteroid type contains carbon and organic compounds, as well as clays and other hydrated minerals. These bodies contain significant amounts of water (up to 20 weight percent). Finding a water-rich NEO would create a logistics depot of immense potential value.

A key advantage of asteroids as a resource is a drawback as an operational environment: They have extremely low surface gravity. Getting into and out of the Moon's gravity well requires a change in velocity of about 2,380 meters per second each way; to do the same for a typical asteroid requires only a few meters per second. This means that a payload launched from an asteroid rather than the Moon saves almost 5 kilometers per second in delta-v, a substantial amount of energy. From the perspective of energy accessibility, the asteroids beat the Moon as a source of materials.

Yet there remains the challenge of working in very low gravity, as well as other difficulties that exist in mining and using asteroidal, as opposed to lunar, resources. First is the nature of the feedstock or "ore." Water at the poles of the Moon is not only present in enormous quantity, tens of billions of tons, but is also in a form that can be easily used: ice. Ice can be converted into a liquid for further processing at minimal energy cost; if the icy regolith from the poles is heated to above 0°C, the ice will melt and water can be collected and stored. The water in carbonaceous asteroids is chemically bound in mineral structures. Significant amounts of energy are required to break these chemical bonds to free the water, at least two or three orders of magnitude more energy than to melt ice, depending on the specific mineral phase being processed. So extracting water from an asteroid (present in quantities of a few percent to maybe a couple of tens of percent) requires significant energy; water ice at the poles of the Moon is present in greater abundance (up to 100 percent in certain polar craters) and is already in a form that is easy to process and use.

The processing of natural materials to extract water has many steps, from the acquisition of the feedstock, to moving the material through the processing stream, to the collection and storage of the derived product. At each stage, we typically separate one component from another; gravity serves this purpose in most industrial processing. A challenge to asteroid resource processing is to devise techniques that do not require gravity, including related phenomena, such as thermal convection, or to create an artificial gravity field to ensure that things move in the right directions. Either approach significantly complicates the resource extraction process.

The great distance from the Earth and poor accessibility of asteroids compared to the Moon works against resource extraction and processing. Human visits to NEOs will be of short duration, and because radio time lags to asteroids are on the order of minutes, direct remote control of processing will not be possible. Robotic systems for asteroid mining must be designed to have a large degree of autonomy. This may become possible but presently we do not have enough information on the nature of asteroidal feedstock to design, or even envision, the use of such robotic equipment. Moreover, even if we did fully understand the nature of the deposit, mining and processing are highly interactive activities on Earth and will be so in space. The slightest anomaly or miscalculation can cause the entire processing stream to break down, and in remote operations, it will be difficult to diagnose and correct the problem and restart it.

The accessibility issue also cuts against asteroidal resources. We cannot go to a given asteroid at will; launch windows open for very short periods and are closed most of the time. This affects not only our access to the asteroid but also shortens the periods when we may depart from the object to return our products to near-Earth space. In contrast, we can go to and from the Moon at any time, and its proximity means that nearly instantaneous remote control and response are possible. The difficulties of remote control for asteroid activities have led some to suggest that we devise a way to "tow" the body into Earth orbit, where it may be disaggregated and processed at our leisure. I shudder to think about being assigned to write the environmental *impact* (if you'll pardon the expression) statement for that activity.

So where does that leave us in relation to space resource access and utilization? Asteroid resource utilization has potential, but given today's technology levels, it has uncertain prospects for success. Asteroids are hard to get to, have short visit times for round-trips, difficult work environments, and uncertain product yields. Asteroids do have low gravity going for them, which is both a blessing and a curse. In contrast, the Moon has the materials we want and in the form that we need. The Moon is close and easily accessible at any time and is amenable to remote operations controlled from Earth, in near-real time. We should go to the Moon first to learn the techniques, difficulties, and technology to conduct planetary resource utilization by manufacturing propellant from lunar water. Nearly every step of this activity, from prospecting and processing to harvesting, will teach us how to mine and process materials from future destinations, on both minor and planetary-sized bodies. Learning how to access and process resources on the Moon is a skill that transfers to any future space destination.

The Moon: Our Next Destination in Space

The Moon is the first extraterrestrial object after leaving Earth orbit and it is a highly desirable place to visit and utilize. Why would we *not* want to explore and use it? Yet, as we have seen, two presidential attempts to return to the Moon in the past twenty-five years have both ended in failure, stifled by bureaucratic process and the continuing siren call of Mars. Other nations clearly see the value of the Moon. Why can't we?

In part, America is the victim of its own early success on the Moon. The Apollo missions and the associated robotic missions that preceded them, were great technical and emotional triumphs. They produced sights and experiences that have yet to be surpassed, even by the technically more challenging (but also more prosaic) flights of the space shuttle and the construction of the ISS. It wet our appetites for more. Because of Apollo, there is a sense that we've been there, and overly anxious explorers don't see a reason to return. This ignorance and quick dismissal about what the Moon has to offer is exploited by space advocates who have other agendas: to quest for life, to step onto new worlds, to build colonies and transform other planets. None of those motivations by themselves have had any better success in generating more—or even adequate—funding for the civil space program. In particular, the constant and recurring obsession with human missions to Mars has kept us from pursuing the more valuable and emphatically achievable near-term goal: a permanent return to the Moon.

Simply put, most people are indifferent to space. This has been true since the beginning of spaceflight, even during the Apollo program.[14] They are neither overly enthusiastic nor hostile to it; they are at best, mildly interested in space, occasionally becoming enthusiastic and patriotic in times of significant accomplishment. For years, space advocates have had the obsessive certainty that if they can impart to the public the same zeal that they feel for Mars or space colonies, or whatever their cause, that they will be showered with *more* money, forever. That hasn't happened and it won't. At best, there will be a modest level of ongoing federal funding—more or less what NASA has received since the end of the Apollo program.

We must craft a program that will endure for decades, a program that makes steady, constant progress and returns tangible benefits with the levels of funding likely to be made available. Our challenge is to work with what we have. Yet, how can we craft a program that aims for big goals, like space settlement or planetary missions, under existing constrained budgets? I have spent the last few years exploring that question, and I believe there is a clear path forward.

7

How? Things We Should Have Been Doing

Although several attempts to revitalize lunar exploration met with partial success, currently, the Moon is not a strategic destination for the United States. A general misunderstanding of the value of the Moon keeps stalling our plans to return. The Apollo program was a successful architecture for getting people to the Moon and that seminal experience still colors many viewpoints on how to approach a lunar return. The Apollo program was the product of historical circumstances born during a specific time and place. While that experience holds many lessons for us, we must resist using it as a guidebook to get back to the Moon.

Questions on how to extend human reach beyond LEO have preoccupied the space community for years, with widely varying opinions on the appropriate steps to take, the order in which to take them, and how to implement the specific technical needs of each phase of human travel in deep space. Although many of these choices are a matter of personal preference, there is a common set of requirements that any trans-LEO architecture must satisfy. In what follows, I will outline some of the basic challenges of human spaceflight, the specific issues confronting travelers beyond LEO, and how these issues can be addressed.[1]

Some Spaceflight Basics

Rocket engines work through combustion. The chemical energy stored in propellant is released and expelled through a nozzle at high velocity. We have many choices—the type of fuel and oxidizer, the engine configuration, fuel flow rates and the geometry of the combustion chamber, as well as the mixing ratios and the nozzle diameter to vary the amounts of power a given rocket engine can generate. Regardless of how we might vary these parameters, we remain fundamentally limited in what we can put into space from the surface of the Earth.

The principal limiting factors in spaceflight are the force of gravity and the amount of energy available for release in the chemical bonds of the propellant. We can do nothing about either of these two factors; they are dictated by nature. At best, we can be clever in our engineering by employing strategies like staging, and by varying the types of materials used to make structures. But varying these parameters work only at the margins, not at the fundamentals. Those fundamentals are described by something called the rocket equation, first formulated in 1903 by the Russian "Father of Astronautics," Konstantin Eduardovich Tsiolkovsky. The rocket equation essentially says that for chemical fuels, a rocket must consist of about 80–99 percent propellant by mass. This depressing arithmetic informs us that the payload—the useful mass that we want to get into space—can be only a small fraction of the mass of the vehicle.[2]

This simple fact of life, one that astronaut Don Pettit aptly terms "the tyranny of the rocket equation," means that going into space is possible, but difficult and expensive.[3] Typical commercial launch vehicles (CLV) are able to put 2–30 metric tons of payload into low Earth orbit, at a cost of between $30 million and $500 million per launch. These costs must include the necessary infrastructure costs, such as ground support, tracking, and insurance. All the air, water, food, and equipment the crew needs during the mission must be brought up by launch. This manifest is in addition to the mass of the launch vehicle, including its structure, tankage, and avionics.

To achieve orbit, a payload must be launched along a carefully chosen trajectory with a rocket burn of precise magnitude and duration. It must be lifted above the atmosphere so that aerodynamic drag does not slow the payload down to roughly 100 kilometers above the Earth, a point called the Karman line, the boundary between air and space.[4] It must be accelerated to a velocity of about 7.8 kilometers per second; at this speed, the distance traveled by the vehicle per unit time is greater than the magnitude of the curvature of the Earth. When this condition is achieved, the launched object will constantly circle the Earth—it is in orbit. At the altitudes of low Earth orbit (~200–300 km), traces of atmosphere occur, meaning that an orbit will eventually deteriorate over time. Because of atmospheric drag, a satellite in LEO eventually will reenter the atmosphere. To alleviate this problem, satellites carry small amounts of fuel that are burned in small rockets, thrusters fired in controlled bursts, to maintain its orbit.

To go beyond LEO to high geosynchronous orbit (36,000 kilometers above Earth), an L-point, or to the Moon or planets (see figure 6.1), additional velocity (positive delta-v) must be imparted to the spacecraft by a rocket burn in the current direction of travel. An engine burn requires propellant, but existing launch vehicles reach LEO with empty fuel tanks. The only way around this problem is to include the fuel needed for trans-LEO travel as part of the payload, which further reduces the remaining fraction available for useful payload, or refuel the upper stage from a stored supply already in LEO. The first method requires the development of something called a heavy-lift launch vehicle (HLV). Such a rocket's specific size is not rigorously defined, but typically, an HLV is able to put 50 to 100 metric tons or more into LEO. An example of an HLV is the Saturn V of the Apollo era, which could launch 116 metric tons into space. The Saturn V was the biggest launch vehicle the United States ever built and was sized specifically for the requirements of a human mission to the Moon, which included the Saturn IV-B upper stage, the Lunar and Command-Service

Modules, and the liquid hydrogen-oxygen fuel needed to send the entire structure to the Moon.

The alternative technique for travel beyond LEO is to store propellant at a depot in space, then refuel the departure stage from that source.[5] The idea of propellant depots in low Earth orbit has drawn a lot of attention, especially from many armchair engineers who have never actually flown a mission beyond LEO. Although this sounds like a good idea—indeed, it is a spacefaring skill that we must eventually master—the hidden assumption of the depot concept is that we possess the capability to launch the propellant "cheaply" from Earth, usually via some magically inexpensive "commercial" source, and store it in orbit. This is a simplified account of the depot concept; many other complex variables must be considered such as propellant boil-off, transfer techniques, management of the arrivals and departures of the tankers, and manifesting the facilities and timings of each launch of propellant cargo. Propellant depots are something that we eventually will take advantage of, particularly when we are ready to export propellant from the lunar surface. For the moment, the use of depots is invoked primarily as a substitute for a heavy lift launch vehicle. In the future, once we begin to produce and export propellant from the Moon, depots will be essential for supplying the vehicles of cislunar and planetary spaceflight.

A benefit Earth provides is that we can decelerate returning spacecraft using atmospheric friction dissipated as heat for braking, thus eliminating the need for propellant to slow down a returning spacecraft, thus making practical spaceflight possible. All returning human missions to date have used this technique, called aerothermal entry. A variant of this concept is aerobraking, in which a vehicle does not actually land, but uses the atmosphere to slow down enough to enter orbit around a planet. Although not used yet on human missions, this approach has been used on some robotic spacecraft sent to orbit Venus and Mars.[6] As part of a system that can be reused and expanded, aerobraking is another skill that must be mastered in order to develop a permanent space transportation system.

The rocket equation dictates that while travel to LEO is difficult, travel beyond it becomes increasingly more so. Although the actual numbers vary depending on the propulsion system and its fuel, putting a single kilogram in lunar orbit requires about five kilograms in LEO, while landing a single kilogram on the lunar surface requires about seven kilograms in LEO, most of which is propellant. A system that enables routine access to cislunar space—the volume of space between Earth and Moon, including the lunar surface—could be established by setting up staging areas where the intermediate travel segments to the varying levels of cislunar space might be launched. Examples of such staging areas include LEO—the ISS is one possibility, GEO, a useful location to access communications and weather satellites; the Lagrangian (libration) points, of which L-1 and L-2 are often mentioned; and lunar orbit, with a variety of possibilities. At these locations, different spacecraft and pieces may be assembled to travel to the next location; they would also be locales for the establishment of propellant depots. A network of transportation nodes will enable constant and routine flight throughout cislunar space.

The tyranny of the rocket equation makes spaceflight difficult, and therefore expensive. It is possible to save some money by using clever engineering and some specialized tricks, but typically, such approaches only nibble around the margins and do not take big bites out of the core cost. This reality—the limiting arithmetic of spaceflight—cannot be addressed with finality as long as we haul everything we need up from the bottom of the deepest gravity well in the inner solar system. We will break loose from our tether once we learn how to create new capabilities by provisioning ourselves from what we find in space.

Launch Vehicle Options

After thirty years of service, the space shuttle was retired in 2011. Many observers regarded the shuttle as unsafe and inefficient, but although fourteen people died in two vehicle failures, 341 people safely made the trip to and from LEO, some taking multiple voyages. Moreover, the failure of two flights, *Challenger* and *Columbia*, out of a total of 135 flights, gives the space shuttle a 98.5 percent success rate, one of the best in the history of spaceflight. No one considers the loss of human life, even those who choose to challenge the limits, as anything but tragic, but each loss of vehicle and crew led to safer subsequent flights. At the end of the program, the shuttle was operating about as safely as any Earth-to-LEO transportation system could.

An enduring problem with the shuttle system was the amount of time and effort needed to refurbish it after each flight. Of the shuttle stack, only the external tank was discarded; all other pieces were recovered and reused. The solid rocket motor segments were simply refilled with propellant. However, the orbiter required many man-hours of work to prepare for launch, especially the silica tiles used to protect the vehicle during the searing heat of reentry. Copious labor-intensive work on the thermal protection system vacuumed up money (during its years of operation, shuttle operations took up the major fraction of the human spaceflight budget). For this reason, some critics consider the shuttle a policy failure, in that it did not make spaceflight to and from LEO "cheap," even though that was never one of its design goals.[7] However, a better way to look at the shuttle is that its goal as a vehicle was to make spaceflight "routine"—and it did. Moreover, the size and design of the shuttle gave it some unique capabilities, some not available on any planned future American spacecraft.

Now that the space shuttle is a historical relic, we are essentially in the beginning days of a new human spaceflight system. Currently under development are the remnants of Project Constellation: the Orion spacecraft and the Space Launch System (SLS).[8] Orion can be configured to carry up to six passengers, four for cislunar flights, and has the capability to reside in space for about three weeks. This duration is adequate for almost all cislunar missions, but for missions beyond the Moon to, say, Mars or an asteroid, Orion will need additional modules for habitation, planetary landing, and other functions. In essence, Orion is only a single piece of a trans-LEO spacecraft system. Moreover, the design of the Orion command module is not conducive to satellite servicing, landing on a planetary object or extensive EVA; since it has no airlock, the entire spacecraft must be depressurized before crewmembers can egress.

The new rocket under development is a heavy lift vehicle (HLV), the Space Launch System (SLS). The SLS is built with pieces derived from the retired shuttle system, including its engines (modified shuttle main engines, burning LOX-hydrogen fuel), its solid rocket motors, and its central core tankage. In its basic form, the SLS can put about 70 metric tons into LEO; there are plans to increase that capacity, first to about 100 tons and ultimately to 130 tons. Depending upon the architecture, this core 70 ton payload capacity is adequate for most lunar missions. The largest variant of the SLS is scaled for human Mars missions staged completely from Earth. In such a case, eight to twelve separate launches are needed to assemble the 500+ ton Mars spacecraft in Earth orbit.

The principal advantage of an HLV is that the number of launches needed to conduct a mission is minimized. Each launch has a finite probability for failure, which is multiplied by the total number of launches. An architecture that uses a smaller LV has greater total risk, even though the impact of the loss of a single vehicle is lessened. Moreover, because ground infrastructure tends to be limited for most launch vehicle systems, the management of resources such as personnel, timing, and processing streams becomes a significant factor in conducting trans-LEO missions, depending on how far the mission is to go and how difficult it might be, lunar missions being easiest and Mars missions being the most demanding. Cost is also a consideration; HLVs tend to have greater economies of scale in terms of dollars per kilogram delivered, but they carry higher initial development and operating costs.

A return to the Moon can be accomplished using smaller launch vehicles and propellant depots. Several different architectures, with varying degrees of realism, have been developed to accomplish such a mission. In all cases, technical difficulties need to be solved before a viable transportation system is developed. The biggest unknowns are associated with the building and operation of propellant depots; delivery, storage, and transfer of propellant are technical issues that have yet to be demonstrated. These are particularly acute with cryogenic propellants (liquid oxygen and hydrogen) that have extremely low boiling points; some propellant will gradually be lost through sublimation regardless of how thermally well insulated the storage tanks are at the depot. One way to mitigate this loss is to store the propellant as water and crack it into its elemental form just prior to use. Such a strategy requires building a substantial infrastructure at the depot, including large solar arrays to generate high levels of electrical power to crack the water and processing facilities to capture and freeze the dissociated gases. This approach makes depots much more complicated facilities than simple storage tanks in orbit. It is possible to provision depots with storable propellants, that is, noncryogens that are much less susceptible to boil-off loss. But storables such as hydrazine and nitrogen tetroxide have much less specific impulse (total energy) when used, and the depots would not be configured to accept and use lunar-produced propellant in the future. The technical complexities associated with cryogenic oxygen-hydrogen depots make their development a protracted effort but after their establishment, one that provides the most extensibility, flexibility, and utility for spacefaring in the long run.

Retaining cryogenic propellant with minimal boil-off is an important issue, but in addition, transfer of supercold liquids in microgravity is a procedure that has yet to be attempted. Various complications of depot configuration are needed to enable the transfer of liquids in orbit, including the use of ullage by inert gases such as pressurized helium or by spinning the depot to generate small accelerations, causing liquids to move in a predictable direction. All of these systems need to be space-certified, meaning that moving parts must be designed for operation in extreme thermal and vacuum environments, which is costly. Presumably, much of the necessary operational work can be automated, but as we have not yet demonstrated the technologies needed for a space-based cryogenic depot, we cannot even begin to design the needed robotic systems. Human intervention and adjustment of the depot machinery probably will be necessary. Most likely, the earliest space-based propellant depots will be human-tended by technical necessity, not by programmatic requirement.

A variety of expendable launch vehicles are now or soon will be available that could implement a propellant depot-based architecture. The largest

ELV available commercially is the Delta IV-Heavy,[9] which can lift 26 metric tons to LEO. Use of such an LV could conduct a lunar surface mission with three launches. Smaller ELVs such as the Atlas 551 (21 tons) or Falcon 9 (11 tons) would require many more launches to stage such a mission. The proposed Falcon Heavy launch vehicle by SpaceX would consist of three strapped-together Falcon 9 vehicles with cross-fed engines.[10] It remains to be seen whether this proposed rocket, with twenty-seven engines burning simultaneously on liftoff, will work and whether it will be fiscally viable as a commercial launch system. If the Falcon Heavy delivers as advertised, it could place about 50 metric tons into LEO, enough to conduct a lunar mission with two launches.

There are many ways to skin the cat of trans-LEO human spaceflight. NASA is currently building a heavy lift vehicle that will enable human missions to the lunar surface in its basic, core configuration (70 tons), so the establishment of propellant depots in LEO is not an immediate necessity. However, because one of the principal goals of a return to the Moon is to learn how to use its resources, establishing a cryogenic propellant depot is an essential piece of a complete system designed to use lunar propellant to fuel space transportation. We will have to address and solve these various technical problems sooner or later, so we might as well build and learn how to operate such a system now.

A Lunar Return Architecture: Leading with Robots

Several attempts to establish human presence on the Moon were abandoned after they foundered on fiscal and political shoals. While there are many reasons for this history, one of the principal ones is the continued and repeated attempt over the last thirty years to re-create the Apollo experience. Apollo, one of NASA's finest accomplishments, took America from essentially zero spaceflight capability to the surface of the Moon in eight years. Unfortunately, this success led the agency to conclude that making leaps in technology and capability through the appropriation and expenditure of massive amounts of federal money was the only viable path to space success. Such an eventuality is extremely unlikely to reoccur. For the foreseeable future, the civil space program will probably be restricted to funding levels of less than one percent of the federal budget, and perhaps much less than that.

Given these restraints, is a trans-LEO human spaceflight program even possible? I believe it is, but we must design an approach that spends money carefully, invests in lasting infrastructure, and uses the resources of space to create new capability. Over the last decade, new data obtained for the Moon have shown that there is abundant water ice at the lunar poles. Moreover, this ice is proximate to locations that receive near-constant solar illumination. These two facts allow us to envision both a location and an activity in cislunar space where an off-planet foothold for humanity could be established. The development of an architecture that works under these constraints and achieves the objectives described above was a joint effort by myself and Tony Lavoie, an engineer from NASA–Marshall Space Flight Center with whom I worked closely on the Lunar Architecture Team in 2006.[11] It should be emphasized that this plan is *flexible*; many aspects of it can be changed to accommodate evolving circumstances, resources, and prevailing societal and political conditions. It is offered as an example of what is possible and not as a detailed master plan that must be followed to the letter.

The *mission statement* of lunar return is "to learn how to live and work productively on another world." We do this by using the material and energy resources of the lunar surface to create a sustained presence there. Specifically, our goal is to harvest the abundant water ice present at the lunar poles with the objective of making consumables for human residence on the lunar surface and propellant for access to and from the Moon and for eventual export to support activities in cislunar space. Initially, the architecture focuses on water production because propellant—in this case, hydrogen and oxygen—is by far the major fraction of vehicle mass and the most significant factor for the cost of human missions. The availability of lunar consumables and propellant allow us to routinely access all the levels of cislunar space, where our economic, national security, and scientific satellite assets reside.

The *objective* of lunar return defines our architecture: we stay in one place to build up capabilities and infrastructure in order to stay longer and create more. Thus, we build an outpost; we do not conduct sortie missions to a variety of landings sites all over the Moon. We go to the poles for three reasons: (1) near-permanent sunlight near the poles permits almost constant generation of electrical power from photovoltaics, obviating the need for a nuclear reactor to survive the fourteen-day lunar night; (2) these quasi-permanent lit zones are thermally benign compared to equatorial regions (Apollo sites), being illuminated at grazing solar incidence angles, and thus greatly reduce the passive thermal loading from the hot lunar surface; (3) the permanently dark areas near the poles contain significant quantities of volatile substances, including hundreds of millions of tons of water ice.

The return to the Moon is accomplished gradually and in stages, making use of existing assets both on Earth and in space. Early missions send robotic machines that are controlled by operators on the Earth. The short radio time-delay permits near instantaneous response to teleoperations, a virtue provided by the Moon's proximity to Earth. An important attribute of this architecture is flexibility. We build infrastructure incrementally with small pieces on the Moon, operated as a single large, distributed system. The individual robotic machines have high-definition, stereo real-time imaging, anthropomorphic manipulation capabilities, and possess fingerlike end-effectors. The intent is to give the robotic teleoperators the sense of being physically present and working on the Moon. These surface facilities are emplaced and operated as opportunity and capability permit. Because there are many small pieces and segments involved in a distributed system, an incremental approach enables a broader participation in lunar return by international and commercial partners than was possible under previous architectures.

The advantage of using smaller units for robotic machines is that they can be either grouped together and launched on one large HLV or launched separately on multiple, smaller ELVs. Such flexibility allows us to create a foothold on the Moon irrespective of budgetary fluctuations.

Commonality occurs at the component level, with common cryogenic engines, valves, avionics boxes, landing subsystems, filters, and connectors to allow maximum use and reuse of the assets that are landed on the surface. The goal is to create a remotely operated, robotic water mining station on the Moon. People arrive at the outpost late in the plan to cannibalize common parts, fix problems, conduct periodic maintenance, upgrade soft goods, seals, valve packing, inspect equipment for wear, and perform certain logistical and developmental functions that humans do best.

Phase I: Resource Prospecting. We first launch a series of small robotic spacecraft to: (1) emplace critical communications and navigational assets; (2) prospect the polar regions to identify suitable sites for resource mining and processing; and (3) demonstrate the steps necessary to find, extract, process and store water and its derivative products. The poles of the Moon have intermittent visibility with the Earth, which creates problems for operations that depend on constant, data-intensive communications between Earth and the Moon. Moreover, knowledge of precise locations on the Moon is difficult to determine and transit to and from specific points requires high-quality maps and navigational aids. To resolve both these needs, we establish a small constellation of satellites that serves as a communications relay system, providing near-constant contact between Earth and the various spacecraft around and on the Moon, as well as a lunar GPS system which provides detailed positional information, both on the lunar surface and in cislunar space. This system can be implemented with a constellation of small (~250 kg) satellites in polar orbits (apolune ~2,000 km) around the Moon. Such a system must be able to provide high bandwidth (several tens to hundreds of Mbps) for communications and positional accuracy (within 100 m) necessary to support transit and navigation around the lunar poles.

Two rovers will be sent to each lunar pole to explore the light and dark areas and to characterize the physical and chemical nature of the ice deposits. We must understand how polar ice varies in concentration horizontally and vertically, learn about the geotechnical properties of polar soils, and pinpoint location and access to mining prospects. The rovers will begin the long-term task of prospecting for lunar ice deposits so that we may select the outpost site near high-concentration deposits of water. In addition to polar ice, we must also understand the locations and variability of sunlit areas, as well as the dust, surface electrical-charging and plasma environment.

The rovers weigh about five hundred kilograms and carry instrumentation to measure the physical and chemical nature of the polar ice. In addition, they will excavate (via scoop, mole, and/or drill) and store small amounts of ice/soil feedstock for transport to resource demonstration experiments mounted on the fixed lander in the permanent sunlight. Because the rover must journey into and out of the permanent darkness repeatedly, it cannot rely solely on solar arrays to generate its electrical power. Power has to be provided by a continuously operating system, such as a radioisotope thermal generator (RTG).[12] Possible nonnuclear alternatives include rechargeable batteries or a regenerative fuel cell (RFC).

During this phase, a propellant depot will be placed in a 400 km Earth orbit to fuel future spacecraft going to the Moon. Initially, the depot will be supplied by water delivered from Earth, but later from the Moon via space tugs. At the depot, water will be converted into gaseous hydrogen and oxygen and then will be liquefied and stored. This depot will fuel a robotic heavy lander with roughly eight metric tons of propellant and must be flexible enough to control its attitude in many configurations during both the absence and presence of docked vehicles. Using the depot to fuel a large lander increases our potential landed mass on the Moon by more than a factor of two. The depot will be supplied initially with water by commercial launch vendors, which can begin immediately after orbit emplacement and checkout. If no commercial providers emerge, separate NASA missions can supply the depot with water.

Phase II: Resource Mining, Processing, and Production. The next phase moves from resource prospecting and exploration to water production. The initial processing approach will be to excavate ice-laden soil, heat it to vaporize the ice, collect the vapor, and store it for later use. It is possible that other, more efficient mining schemes, such as some type of in-place extraction, may emerge that do not require soil excavation. For now, the most conservative approach, one that we know will work, is to use heat to drive the water from the soil. The process of soil heating has the advantage of being able to use either electrical power or passive solar thermal energy to generate heat for the processing of the feedstock.

During this phase, we incrementally add excavators, dump haulers, soil processors, and storage tanks to obtain, haul, and store the water.[13] Landers carrying large solar arrays generate electricity at the permanently illuminated (> 80%) sites; robotic equipment can periodically connect to these power stations to recharge their batteries (figure 7.1). Our immediate goals are to learn how to remotely operate these machines and begin to produce and store water for eventual use when people arrive. Processed water is easily stored in the permanently shadowed areas. During this phase we also land electrolysis units to begin cracking water into its component gases, making the cryogens, and storing the liquid propellant. Because we are developing an operational cadence as we go,

Figure 7.1. Artist's rendering of robotic lander approaching surface. Previous lander has deployed solar arrays that rotate on vertical axis, to track Sun near pole. These robotic systems can begin the work of resource processing at the lunar poles.

it might take several months to get into a smooth rhythm that maximizes the rates of propellant production. Large unknowns that must be resolved include transit time between the mining and propellant production site, thermal profiles, power profiles, and the lifetime of machine parts. We make constant, steady progress, learning how to crawl before we try to walk.

Equipment used in this phase includes excavation rovers, processors, and power units, each on the order of 1,200–1,500 kilograms. Power stations are rolled solar arrays that when deployed, are gimbaled about a vertical axis to track the course of the Sun over a lunar day. Each array generates about 25 kW. Multiple power stations can be arranged and operated together to provide the power needs of the robotic equipment and, ultimately, the outpost. During this phase, we begin to investigate the making of roads and cleared work areas by microwave sintering of regolith. Many areas near the outpost site, particularly around the power stations, will get heavy repeat traffic and keeping scattered dust to a minimum is necessary for thermal control and to maximize the equipment lifetime.

Phase III: Outpost Infrastructure Emplacement and Assembly. The next phase will bring pieces of the outpost and prepare its site, emplace critical infrastructure for power generation and thermal control, and begin to construct the lunar surface transportation hub, which will receive and service the reusable robotic and human landers that make up our cislunar transportation system. Additional robotic assets are added, including upgrading the surface mining and processing equipment, replacing damaged items, and expanding the capacity of processing (figure 7.2). Our goal in this phase of development is to increase the output of water in order to support the arrival of human crews on the Moon.

Propellant is needed on the lunar surface to refuel the robotic and human landers that travel to and from the Moon. Returning cargo landers can carry the exported product as water or as propellant. Both options may be necessary, since propellant will be needed in the vicinity of the Moon to refuel transfer stages, but water delivered to low Earth orbit can be cracked and frozen there just as efficiently as on the lunar surface. Included in the

Figure 7.2. Artist's rendering of robotic bulldozers digging ice-laden regolith as feedstock for water extraction processing. The proximity of Earth allows us to teleoperate robotic machines on the Moon and begin resource processing prior to human arrival.

power budget is the energy required for propellant liquefaction, which removes a large amount of heat from the fluid. Minimizing the boil-off of the volatile cryogens is a recognized technical challenge and will be addressed via selected technology development early in this campaign.

The first heavy cargo mission will bring the logistical pieces and power capability necessary to support human habitation for the initial stay on the lunar surface. Part of this cargo includes additional power generation capability to power the human habitat arriving later. The initial cargo complement would probably not include enough battery power to weather an eclipse, but it is expected that this capability would arrive by the third cargo mission. Part of this complement would be supplementary equipment needed to attach to the habitat or otherwise make it usable, such as leveling equipment, high priority spares, filters, thermal shields, various pieces of support equipment, lifting equipment, mobile pallets, EVA suit components, and logistics supplies, including a method to transfer the crew to the habitat in the form of a tunnel/airlock so that the mass of the human lander can be minimized. Included is a small, pressurized human rover

(4.5 tons) to interface between the lander and the habitat to allow shirt-sleeve ingress, as well as local mobility to access deployed equipment.

The second heavy cargo mission brings the human habitat to the Moon. While it is envisioned that the habitable areas at the outpost ultimately will be significantly larger than a single 12-ton module, initial needs are to have sufficient habitable volume to support two to four crewmembers for a month. Included in either this or the previous mission payload are radiators and heat rejection equipment, as well as a fully operational environmental control and life support system.

Phase IV: Human Lunar Return. During this phase, we prepare the site, emplace the elements, and connect all the pieces to create a ready-to-use outpost. Those pieces include power and thermal control systems, habitats, workshops, landing pads, roads, and other facilities. Remotely operated robotic machines assemble this entire complex before people arrive. The outpost is "human-tended" and supports a crew of four for biannual visits of several weeks duration. During these periods, the crew repairs, services and operates the previously emplaced robotic assets. In addition, some of the crew will conduct local geological exploration and other science-related tasks. By the time the first crew arrives, the outpost will be producing about 150 tons of water per year, enough to completely supply the lunar transportation system with propellant.

The lander for these human missions is a smaller, LM-class vehicle (~30 ton) rather than a lander similar to the Constellation *Altair* vehicle (~50 ton). Its primary mission is to transport crew to and from the lunar surface and does not contain significant life support capability, since the crew will live in previously emplaced surface habitats while they are on the Moon. This lunar taxi becomes a permanent part of the cislunar transportation system. It is reusable and refuelable with lunar-produced propellant and can be stored either on the lunar surface or at the cislunar transport node. Because of its similarity in size and functionality to the robotic landers, common components are used so that the parts count for lunar surface mainte-nance can be minimized. Specifically, both landers use a common reus-able cryogenic engine developed in part (or totally) for use by the robotic heavy lander, with both vehicles using a multiple engine complement for reliability and redundancy, as well as cost. Engines will be designed to be serviced or changed out on the Moon, thus maximizing the lifetime of the vehicles in which they reside.

With refueling at the LEO depot, a cargo variant of the human lander launched on a HLV can deliver 12 tons of payload to the lunar surface. Once on the Moon, it will be cannibalized and used for parts. The lander has a dry mass of 8,300 kg and is launched from the LEO station using a Cislunar Transfer Stage (CTS), which requires about 60,000 kg of cryogenic propellant to take the lander to the Moon. Initially, the CTS will be used and discarded, but once lunar propellant production is up and running, we can reuse this element by rendezvousing in low lunar orbit with the cislunar depot. This architecture does not presume full success with extracting lunar resources, except for refueling for human Earth return. As the concept matures and our understanding of the logistics, cost, and sustainability of this approach solidifies, the lunar refueling process can expand significantly, as much as the demand will allow, to include the incorporation of the cargo landers.

Phase V and beyond: Human Habitation of the Moon. Once the outpost has been established, initial human occupancy will consist of periodic visits designed to explore the local site and to maintain and assure the proper operation of the mining and production equipment. These visits will be interspersed with the landing of additional robotic assets to continually increase the level of production, with the aim of exporting surplus water to cislunar space. Initially, the crew will validate and ensure the propellant and water production chain, including periodic maintenance and optimization of the operations concepts and timelines. With subsequent cargo deliveries, the crew will evaluate production techniques, procedures, technologies, and tools that allow expansion to the next step in utilization (figure 7.3).

Although the concept of lunar resource utilization has been studied for years, many unknowns need to be addressed, starting with basic technologies and technology applications in the lunar environment. Techniques, tools, and extensive physical and metallurgical analysis of the properties of

Figure 7.3. Example of resource processing at a lunar outpost during early phases of operation. A vacuum induction furnace takes metal obtained from the regolith, melts it and pours the liquid into molds to make metallic members for construction. A crane at right is unloading a payload from a robotic lander. Electrical power is provided by solar array landers in the distance at top left.

the final products need to be examined to obtain the best products for as yet undefined applications. Research in this technology is vitally important to extending human reach in space, although habitat upkeep and propellant supply chain management has higher priority. A broad ISRU material investigation lends itself well to both international participation and commercial development. Because no single strategy or technology or method works for every application, research can be divided into discrete investigations. Toward that end, on one of the cargo missions, a materials processing laboratory is delivered. Next in priority for crew time is data on biological interaction and plant growth in lunar gravity. These investigations will examine the vitality, reaction, and long-term logistical needs for developing local food production to sustain human habitation of the Moon.

At this stage, we may begin to recoup our investment in the outpost. Several possible models for the privatization of water processing may be viable. We anticipate that the federal government will be an early and repeat

customer for lunar water, not only for future NASA missions beyond the
Earth-Moon system but also for the cislunar missions of other agencies,
such as the Department of Defense. Additionally, international customers
will likely emerge. Whether the production facilities become commercial-
ized before or after these markets emerge cannot be easily foreseen at this
stage and in fact, is unimportant. The critical point is that we will be in a
position to industrialize the Moon and cislunar space, a cornerstone in making space part of our economic sphere. We can openly share the technology
developments as well as any undesirable outcomes and pitfalls from our
experience, so that others can leverage what we have learned. This will enable
the commercial sector to take over many lunar activities and services.

The transition to commercial activity may occur early or late in outpost development. Part of NASA's ultimate purpose is to expand and
enhance the nation's commercial and industrial base and this activity is
to be encouraged where possible. However, in contrast to NASA's obsession with devising an "exit strategy" for the Moon, we should instead plan
to participate in lunar development for at least as long as deemed necessary for fully commercial (that is, not government subsidized) providers
to emerge. Because the capabilities we are developing have critical national
strategic importance, the involvement of the federal government is important to ensure continuing access to lunar resources and the capabilities
they provide.

Establishing a permanent foothold on the Moon opens the space frontier
to many different uses. By creating a reusable, extensible cislunar spacefaring system, we build a "transcontinental railroad" in space, connecting
two worlds, Earth and Moon, as well as enabling access to all the points in
between. We will have a system that can access the entire Moon, but more
importantly, we can also routinely access all of our assets within cislunar
space: communications, GPS, weather, remote sensing, and strategic monitoring satellites. These satellites can be serviced, maintained, and replaced
as they age.

I have concentrated on water production at a lunar outpost because
such activity provides the highest leverage through the making of rocket

propellant. However, there are other possibilities to explore, including a paradigm-shifting culture to eventually design all structural elements of space hardware using lunar resources. These activities will spur new commercial space interest, innovation, and investment. This further reduces the mass needed from Earth's logistics train and helps extend human reach deeper into space, along a trajectory that is incremental, methodical, and sustainable within projected budget expectations. Instead of the current design-build-launch-discard paradigm of space operations, we can build extensible, distributed space systems with capabilities much greater than currently possible. Both the space shuttle and the ISS experience demonstrated the value of human construction and servicing of orbital systems. What we have lacked is the ability to access the various systems that orbit the Earth at altitudes much greater than LEO—MEO, GEO, and other locations in cislunar space.

A transportation system that can access cislunar space can also take us to the planets. The assembly and fueling of interplanetary missions is possible using the resources of the Moon. Water produced at the lunar poles can fuel human missions beyond the Earth-Moon system, as well as provide radiation shielding for the crew, thereby greatly reducing the amount of mass launched from the Earth's surface. To give some idea of the leverage this provides, it has been estimated that a chemically propelled Mars mission requires at least roughly one million pounds (about 500 tons) in Earth orbit. Of this mass, more than 80 percent is propellant. Launching such propellant from Earth requires eight to twelve HLV launches at a cost of almost $2 billion each. Such an approach does not establish a true exploration capability. A Mars mission staged from the facilities of a cislunar transport system can use propellant from the Moon to reduce the mass launched from the Earth by a factor of five.

The modular, incremental nature of this architecture facilitates international and commercial participation by allowing their contributions to be easily and seamlessly integrated into the lunar development scenario. Because the outpost is built around the addition of capabilities through the use of small, robotically teleoperated assets, other parties can bring their own pieces to the table as time, availability and capability permit. International partners will be able to contemplate their own human missions to the Moon without the need to develop a heavy-lift vehicle by purchasing lunar fuel for a return trip. Flexibility and the use of incremental pieces make international participation and commercialization in this architecture easier than under the Project Constellation architecture.

These are only the initial steps of a lunar return based on resource utilization. Water is both the easiest and most useful substance that we can extract from the Moon and use to establish a cislunar spacefaring transportation infrastructure. Once established, many different possibilities for the lunar outpost may emerge. It may evolve into a commercial facility that manufactures water, propellant, and other commodities for sale in cislunar space. It could remain a government laboratory, exploring the trade space of resource utilization by experimenting with new processes and products. Alternatively, it might become a scientific research station, supporting detailed surface investigations to understand the planetary and solar history recorded on the Moon. We may decide to internationalize the outpost, creating a common use facility for science, exploration, research, and commercial activity by many countries. By emphasizing resource extraction early, we create opportunities for flexible growth and for the evolution of a wide variety of spaceflight activities.

Schedules, Budgets, Politics, and "Sustainability": Is Any of This Possible?

It is an article of faith in the space community that the US civil space program is woefully under-funded and should receive much more money; some advocate for at least doubling the current NASA budget. Is it really true that the space program does not receive enough money? Certainly, the space program is now funded at a much lower fraction of the federal budget (about 0.3 percent) than was appropriated at the height of the Apollo program (about 4 percent).[14] But at that time (1961–68), NASA had virtually no infrastructure, including laboratories, offices, test stands, launch complexes, and supporting facilities, and little off-the-shelf technology to draw on. Much of the Apollo spending went to these ends and created a supporting network and organizational base that the agency has used and drawn upon for all of its many programs ever since.

As we have seen, previous efforts to return to the Moon were cut short by budgetary shortfalls. In Washington, the estimates for the cost of new programs have a long history of running significantly lower than what things actually and eventually cost. Nonetheless, one problem with talking about money is that the cumulative costs for a multiyear or multidecadal program seem horrendously high.[15] As implemented by the 90 Day Study to support President George H. W. Bush's 1989 Space Exploration Initiative (SEI), the estimated cost was $600 billion; at the time, the agency's yearly budget was a bit more than $10 billion. But that $600 billion number was the total cost of a thirty-year program and included all of the ancillary costs of facilities and overhead. Even though few federal programs could withstand such accounting scrutiny, critics used the $600 billion number as a cudgel to beat the SEI to death. One might stop and consider than in the twenty-five years since SEI was unveiled, the agency has spent about $498 billion (FY 2014) dollars, almost the same gasp-inducing number as that of the 90-Day Study. One might pause and reflect on what that sum has bought us in terms of spacefaring capability over the last two and a half decades.

Rushing in where budgetary angels fear to tread, I now present, in table 7.1, our estimate for the cost of lunar return via the scenario described in this chapter.[16] Tony Lavoie and I assumed federal budget austerity for the indefinite future and used the budget guidelines for the agency assumed by the 2009 Augustine committee as a cost cap; in effect, a maximum of $7 billion (FY2011) constant dollars per year is to be spent on "exploration systems."[17] The Augustine committee concluded that NASA could not return to the Moon under these fiscal constraints and suggested that an additional $3 billion per year would be needed to fulfill the VSE goals. We simply did not believe that conclusion and that disagreement was in part the motivation for writing our paper. We found that by carefully defining our mission objectives up front and using remotely controlled robotic systems on the Moon in the early stages of the program, we could create a permanent resource-processing outpost at one of the poles under fairly tight fiscal restrictions. Our plan costs an aggregate total of $88 billion (FY 2011) constant dollars over the course of about sixteen years. That amount includes the cost of the development of the robotic infrastructure, propellant depots, reusable lunar lander, the CEV, and a medium HLV (70 ton class). It also includes all of the commercial ELV launch costs at the then-quoted rates. At the end of this nominal program, we have an operating, human-tended polar outpost on the Moon that produces 150 tons of water per year.

A critical aspect related to cost is program performance. Any human spaceflight program must show continual progress in order to maintain its level of funding. The best way to accomplish this is to attain significant and recognizable intermediate milestones on a continuing and regular basis. A manager has much more credibility when he can report program accomplishments as he asks for the next increments of funding. Part of the problem with Project Constellation was that its intermediate milestones were too few and far between. In the five years that program ran, the only significant milestone was a launch test in 2009 of the Ares-X, basically a four-segment shuttle solid-rocket booster with a dummy upper stage. When the program was cancelled in 2010, flight tests of Orion into orbit were not scheduled to begin until 2015. Lunar return was over a decade away; the Augustine Committee claimed that it would not occur until after 2030, a completely undocumented assertion but one embraced by the opponents of the VSE, who were eager to terminate the whole effort. The Constellation program's own lack of near-term milestones, accomplished on a regular cadence, allowed this assertion to go unchallenged.

By crafting an incremental program using smaller spacecraft, flight rates are dramatically increased and consequently, many intermediate milestones are achieved early and often. Yet, no capability is lost because the small pieces are operated together as a single, large "system of systems." In addition, a program that is divided into small pieces is more robust in that it can survive budgetary storms with more resilience. Less progress is made

All costs are in millions of US FY 2011 dollars. Cost for each mission and/or mission element shown in [...] Total column at far right; yearly costs shown across bottom row; total program cost at bottom far right. Human mission costs shown in **bold italic**.

Mission	Description	Launch Vehicle	Lander #	Year 1	Year 2	Year 3	Year 4	Year 5	Year 6	Year 7	Year 8	Year 9	Year 10	Year 11	Year 12	Year 13	Year 14	Year 15	Year 16	Total
1	Lunar Communications Satellites	Atlas 401		25	100	175	100													400
2	Characterize Water Deposits	Atlas 551	RML 01, 02	150	350	550	350	200												1600
3	Water Extraction Demo	Atlas 401	RML 03		50	350	250	100												750
4	LEO Fuel Station Phase 1	Atlas 551		100	600	700	550	400	250											2600
5	Water Processor #1	Atlas 551	RHL 001	200	450	600	650	500	420	300										3120
6	Water Tanker	Atlas 401	RML 04				50	150	230	135										565
7	Ore Excavator/Hauler #1	Atlas 551	RML 002			100	150	350	550	440	200									1790
8	Water Electrolysis #1	Atlas 551	RHL 003				100	250	450	350	150									1300
9	Rover Fueling Tanker	Atlas 401	RML 05					50	150	220	145									565
10	LLO Way Station	Atlas 551					50	100	250	400	205	145								1150
11	LEO Fuel Station Phase 2	Atlas 551							150	430	500	300	170							1550
12	Reusable Water Tank Lander	Heavy Lift	RWTL #1					50	300	475	600	510	350	315						2600
13	Human Power and Logistics Cluster	Heavy Lift	CL 01						100	400	600	700	650	550	450					3450
14	Water Electrolysis #2	Atlas 551	RHL 004								100	145	150	205	125					725
15	Human Habitat #1	Heavy Lift	CL 02								50	350	630	870	900	800				3600
16	Human Lander (reusable)	Heavy Lift	HL 01							100	400	500	800	850	900	950				4500
17	***First Human Mission (cost for P/L)***	Heavy Lift										25	100	100	100	175				500
18	Ore Excavator/Hauler #2	Atlas 551	RHL 005								50	100	100	100	175	200				725
19	Water Processor #2	Atlas 551	RHL 006							50	100	100	100	150	100	75				675
20	Water Electrolysis #3	Atlas 551	RHL 007											100	100	325	200			725
21	Human Habitat #2	Heavy Lift	CL 03											60	150	375	375			960
22	***Human Mission 2***	Heavy Lift												50	150	200	100			500
23	***Human Mission 3***	Heavy Lift													50	150	200	100		500
24	***Human Mission 4***	Heavy Lift														50	150	200	100	500
25	Unpressurized ISRU Lab	Heavy Lift	CL 04												50	200	600	600	450	1900
26	***Human Mission 5***	Heavy Lift														50	150	200	100	500
27	***Human Mission 6***	Heavy Lift															50	150	200	400
28	***Human Mission 7***	Heavy Lift															50	150	200	400
29	Water Electrolysis #4	Atlas 551	RHL 008														100	150	250	500
30	Water Processor #3	Atlas 551	RHL 009													100	100	200	200	600
31	Ore Excavator/Hauler #3	Atlas 551	RHL 010															50	200	250
	Heavy Lift Launch Vehicle (includes Ground Systems at KSC)			100	400	1000	1200	1300	1100	1000	1000	1000	1000	1350	1350	1350	1350	1350	1350	17200
	Block 1 CEV			300	600	1200	1200	1000	500											4800
	Block 2 CEV (including TLI Stage)									200	300	425	550	550	500	500	500	500	500	4525
	Cislunar Transfer Stage							100	200	300	400	400	400	400	400	150	150	100	50	3050
	Cargo Lander				1050	225						400	700	200	600	400	450	350	500	3600
	Technology Wedge			2475																3750
	Undefined Mission Wedge															0	1625	2050	2200	5875
	JSC Ops Cost for 2 Human Flts/yr			40	40	40	40	40	120	160	240	240	280	320	400	400	400	400	400	3560
	Architecture Integration			10	10	10	10	10	30	40	60	60	70	80	100	100	100	100	100	890
	Totals per year			3400	3650	4950	4700	4600	4800	5000	5100	5400	6050	6250	6650	6650	6650	6650	6650	87150

Costs include two versions of Orion crew exploration vehicle (CEV), medium-class heavy lift vehicle (HLV, 70 metric ton), technology development funds, and operations costs shown at bottom.

during lean times, but some progress is still made. It is also easier to take advantage of technical breakthroughs and incorporate them into the program because system and vehicle designs are not frozen in place decades ahead of time. As mentioned, an incremental program also facilitates the integration of commercial and international partners, with more "on-ramps" and a lower bar to program entry. Moreover, the possible failure or poor performance of an individual partner has less impact on program progress and viability.

It is difficult to sustain large-scale technological projects over periods of more than a few years. In the history of America, only a few such programs have succeeded and almost all were somehow related to national security concerns. As we shall see, the program to develop a permanent cislunar transportation system is no exception. Although I have described this program as a return to the Moon, it is also a step toward the creation of a permanent spacefaring capability. By building this system, we access on a routine basis, not only the lunar surface, but also all of the other points within cislunar space, where our national scientific, economic, and security assets reside. Other nations are well aware of the security dimensions of this capability, and some, such as China, are actively pursuing the means to possess freedom of access to this theater of operations. A program to create true spacefaring capability has many critical national benefits that transcend politics. A national bipartisan consensus has defended this nation on land, at sea and in the air for more than two hundred years. Can we afford to do less on the new ocean of space?

8

If Not Now, When? If Not Us, Who?

A widespread misconception about the nature and meaning of the Apollo program has greatly contributed to our inability to establish a long-term strategic direction for our civil space program. Essentially, Apollo was a Cold War battle between the United States and the Soviet Union. Once that presidential goal had been achieved and victory declared, we moved on, since you don't keep fighting a battle that you've already won. The norm for America's relationship with the Moon has been on-again, off-again ever since. We raced to the Moon with wild abandon and then left with equal, if not greater, haste. After achieving one of the greatest national technological challenges since the atomic bomb, America departed the Moon with dispatch. The damage caused by misreading the true significance of Apollo is palpable.

Apollo's success dramatically and unquestionably demonstrated that human spaceflight into the solar system is possible, a knowledge permanently engraved in the minds and on the hearts of so many in the space community. Those who made Apollo possible view that era as a lost "golden age" of space exploration, with the ensuing years reduced to the prosaic and mundane tasks of satellite servicing and educational, zero-gravity demonstrations. Ironically, the success of Apollo has contributed to our multidecadal inability to move forward; it has become the crippling, carved-in-stone standard that continues to influence current thinking about our civil space program. Witness the approach of our recent lunar return efforts: Each one followed the well-trod path wherein we devised and planned an Apollo-like program, then, taking our cue from previous efforts, promptly retreated when stacks of cash magically failed to appear on schedule.

It is possible that what's missing in our debate over a return to the Moon is the benefit of a clear-eyed historical perspective, one unique to America. There is no perfect analogy to the space program, but several past events in our nation's history suggest that some general inferences may be drawn. By examining some historical resonances of spaceflight and attempting to draw conclusions about its proper place and significance, perhaps we can discern a more productive, less disruptive path toward space capability.

Lunar Return in Historical Perspective

The United States has undertaken many large-scale, collective projects over the course of its 240-year history, but none more mythologized than the effort to put a man on the Moon ahead of the Soviet Union in the 1960s. The Apollo program has all of the appeal of great drama: A charismatic, martyred president issues a grand and seemingly near-impossible challenge to the nation, one eagerly grasped and accomplished by a can-do country, proving once and for all that the good guys win in the end. It is remarkable how this caricature is so widely accepted. In fact, President Kennedy was a reluctant spacefarer who undertook the race to the Moon only as a way to distract public attention from the less-than-stellar beginning of his first term. Kennedy had ardently asked advisors to come up with some other technical contest, something with a practical benefit that could win over friends and allies in developing nations. The desalination of seawater was his personal favorite.[1]

But space was making headlines in the late 1950s and early 1960s. At the time of Kennedy's declaration in May 1961, it was widely perceived that we were behind the Soviets, not only in space but also in nuclear capability. This was a case where perceptions were more important than reality. It did not matter that the Soviets were woefully behind in the production of missiles that could actually deliver a warhead. They had already humiliated the new president twice, once by thwarting his sponsored invasion of Cuba, at the Bay of Pigs, and then again with the flight of Yuri Gagarin, the first man to orbit the Earth. America was behind in the Cold War and behind in space. Something needed to be done. What followed held momentous consequences for both Cold War rivals, and by extension, all nations.

Apollo was a special product, one of its own time and space; it does not fit into any current recognizable category of circumstances surrounding the creation of a large-scale federal engineering project. But that does not mean that a return to the Moon today is not feasible. Traditionally, such projects are undertaken for economic and/or security concerns. Both motivations are applicable to the problem of lunar return, and as such, answer both of these compelling national needs.

At the time of the 1849 California gold rush, there were only two ways to get to the goldfields. One was a long and tedious sea voyage from the East Coast to San Francisco, with the choice of taking the long route around the tip of South America or traversing the malarial swamps of Panama for a ship transfer midway through the voyage. The other was a hazardous, months-long crawl across the continent through the wilderness of the American interior. The need for a railroad to connect the nation together was a pressing concern. Several visionaries advocated for the construction of a transcontinental railroad to connect California with the rail systems of the east. After a long study and critical review of several routes, a path was selected and its construction was approved by the Congress and signed into law by President Abraham Lincoln. The Pacific Railroad Act of 1862 provided financial incentives and land grants by the federal government for each segment of rail built along the approved route. The Union Pacific and Central Pacific railroads started building inward from their termini (Omaha and Sacramento, respectively) and converged at Promontory Point, Utah, in 1869. The teams symbolized the linking together of the nation by the transcontinental railroad with the hammering of a final golden spike. Now both coasts were accessible, and the continental interior was open to migration, development, and settlement.

Some believe that such an approach is a possible model for the development of space. In place of a government-run space program, government provides a series of incentives and grants whereby private companies are induced to create the necessary spacefaring infrastructure that will see an economic expansion similar to that brought about 150 years earlier by the building of the transcontinental railroad. This analogy is not perfectly aligned with historical realities; in the 1860s, an extensive rail transportation infrastructure already existed, largely capitalized by the private sector. Comparable assets for spaceflight consist of commercial launch suppliers, but they are both less extensive and have narrower and smaller markets in the field of space than did the railroads of the nineteenth century in the field of passenger and cargo transport. Because spaceflight is more difficult and more dangerous than rail travel, the overall volume of traffic—and thus revenue—is much lower, which depresses capital investment. New Space advocates sometimes cite the US Post Office's airmail service of the 1920s as a good business model. Although the Post Office contracted with private air companies to carry the mail, in this case, a large market (the US Mail) already existed—what was being purchased was faster delivery. That commodity is not nearly as desirable in the field of spaceflight, where timeliness is less critical than assured delivery and reliability.

One historical parallel does compare closely to an ambitious space goal in terms of resources needed: the development of the atomic bomb.[2] The largest technological-scientific effort ever undertaken, the Manhattan Project engaged some of the finest scientific and engineering minds in the country. Billions of dollars were spent developing a deliverable bomb, whose feasibility was uncertain when work began. The driving imperative was national survival, always a guarantee for continued funding. The concern was that Germany was actively working on atomic weapons, a supposition later found to be incorrect. In any event, the Manhattan Project was the largest, most difficult technical project ever attempted. Its success led to the idea that government-funded research in science and technology could serve national aims, a lesson subsequently applied to the waging of the Cold War, of which Apollo was one part. The 50-year struggle against the Soviet Union led to the creation of a science-technology industrial sector upon which we drew heavily during the Apollo program. The systematic dismantling of that sector in the 1990s after the fall of the Soviet Union means that techno-industrial base is gone, making progress in space more difficult to achieve today.

Various large-scale construction projects, completed over the years, offer useful lessons in undertaking future national engineering efforts. The United States took over French efforts to build the Panama Canal in 1904, completing it a decade later in 1914. Both engineering expertise and capital investment were judiciously applied to the problems posed by the canal, which revolutionized seafaring and world trade. Noted at the time was the significant national security aspect of canal building; the Panama Canal enabled the United States Navy to easily move ships between the Atlantic and Pacific oceans, creating a responsive force multiplier that turned out to be critical during two world wars. The Interstate Highway System, proposed by President Dwight D. Eisenhower and inspired by the German Autobahn, created a new automobile-based national transportation infrastructure. Its ostensible purpose was to provide a network of roads to serve the needs of national defense, but in addition, its creation has been responsible for expanding national economic activity by trillions of dollars. Thus, large-scale government programs enabled us to move farther afield, to generate wealth and prosperity, and to secure our national defense.

In past efforts, the federal government has led where the private sector has been unable or unwilling. Because spaceflight is inherently a difficult undertaking, one requiring billions of dollars in capital investment, private spaceflight, to date, has focused primarily on the existing satellite launch market. But unlike early aviation, there is no preexisting "air post" service market driving the development of a new private transportation sector. Much hope is currently invested in the envisioned but unfulfilled potential of space tourism as an emerging market. Despite cash awards and other incentives, substantial private human spaceflight remains, for the most part, cost- and market-prohibitive. Potential possibilities for space commerce have been identified in the communications, energy and construction sectors. What is missing is the ability to move cargo and people routinely throughout cislunar space.

The rationale for space development articulated in 2006 by former Presidential Science Advisor John Marburger calls for space to become part of our economic sphere.[3] In part, we have already started to reach that goal, as evidenced by existing commercial markets for satellite communications and remote sensing data. Due to degradation, space assets that reside in orbits above LEO need periodic replacement, along with an occasional upgrade in technical capability. If we could reach these high geosynchronous orbits, satellites could be serviced and maintained. Additionally, we could assemble large distributed systems in GEO using people and robots working together. This approach was documented to be of great value during the 30-year history of the space shuttle program, when astronauts serviced and maintained satellites like the Hubble Space Telescope and built the International Space Station from small modules fabricated and launched separately.

A permanent presence in cislunar space serves many national economic goals. But how is this best achieved? What are the roles of government and the private sector in the development of space? Most important, how can we devise a civil space program that serves the most needs, in the most efficient manner?

The Geopolitical Value of the Moon and Cislunar Space

Modern power projection is possible only through the deployment and use of space-based assets. Air, land, and sea forces all depend on satellites for communications, navigation, and intelligence. Without them, our ability to make our way about in the world would be severely compromised. Satellites are physically very vulnerable. One need not collide with one to disable it—snapping an antenna or cutting a cable to its solar array can turn a billion-dollar satellite into a worthless piece of orbiting space debris. It is essential to protect our national satellite assets, both to safeguard our investment and, more importantly, to assure that they will function at a moment's notice.

Some in the New Space community take a libertarian view of space development. They suppose that government—in the form of NASA, the agency given responsibility for civil spaceflight—is an impediment that creates more problems than solutions. However, a clearly defined constitutional role of the federal government is to provide for the common defense; this includes maintaining the territorial integrity of the United States and the protection of legal and economic interests of American citizens abroad. As more American commercial entities venture into the zones beyond low Earth orbit, their activities and interests become part of the defense obligations of the US government. Thus, it is not merely appropriate but essential that the federal government maintains a visible profile and role in cislunar space.

Government activities in space should consist of those actions designed to protect American interests. This protection requires the projection of national power as needed and appropriate, as well as the establishment of a legal environment whereby individual and corporate rights and obligations are observed and defended. Such a portfolio of activities requires the physical presence of government. If the government is not present in such a theater, how can it enforce its regulatory and legal decisions? One possibility is through asset seizure on Earth, but such a technique will only stifle, not encourage, space development. Just as the US Navy defends freedom of the seas and the commerce of all nations, an American presence in cislunar space will likewise defend and assure freedom of access and commerce there.

The American civil space program was originally established to conduct research into the techniques and possible beneficial uses of spaceflight. That mandate is declared in the Space Act of 1958, subsequently amended many times.[4] The act outlines the role of the federal government in space and consists of nine basic objectives, including the attainment of scientific knowledge, the development of space technology and flight systems, and international leadership and cooperation. The Space Act effectively authorizes NASA to conduct virtually all imaginable activities in space, including the creation of new spaceflight capabilities.

To achieve a paradigm shift in spaceflight, we must understand how we can use lunar and space resources to create new capabilities and how difficult such activity might be. Despite decades of academic study, no one has demonstrated resource extraction on the Moon. There is nothing in the physics and chemistry of the materials of the Moon that suggests it is not possible; we simply do not know what practical problems might arise. This is why resource utilization is an appropriate goal for the federal space program. As a high-risk engineering research and development project, it is difficult for the private sector to raise the necessary capital to understand the magnitude of the problem from the perspective of an end-to-end system. The original VSE was conceived to let NASA answer these questions and to begin the process of creating a permanent cislunar transportation infrastructure. As an engineering research and development project with uncertain prospects for success, such an effort is entirely appropriate for the federal government to undertake. The results of this project could lead to the creation of new markets and wealth, as the private sector will then possess the strategic knowledge necessary to take advantage of the economic opportunities provided by cislunar space development.

China and America: A New Space Race?

Just as America is standing down from space leadership, China is stepping up its program to send people to the Moon. This circumstance has reawakened a long-standing debate about the geopolitical aspects of space travel and with it, some questions. Are we in a race back to the Moon? Should we be? And if there is a "space race" today, what do we mean by the term? Is it a race of military dimensions, or is such thinking an artifact of the Cold War? What are the implications of a new space race?

Many who work in the space business purport to be unimpressed by the idea that China is going to the Moon, even publicly inviting them to waste money on such a stunt. "No big deal" seems to be the attitude—after all, America did that more than 40 years ago. NASA Administrator Charles

Bolden professes to be unmoved by the possible future presence of a Chinese flag on the Moon, having noted that there are already six American flags there. It should be further noted that 40 years of exposure to solar ultraviolet radiation has probably bleached and faded the red, white, and blue into a dull white.[5]

Although it is not currently fashionable in this country to think about national interests and the competition of nations in space, others do not labor under these restrictions. Our current human spaceflight effort, the International Space Station (ISS), has shown us both the benefits and drawbacks of cooperative projects. Currently, we do not have the ability to send crews to and from the ISS. But that's not a problem; the Russians have graciously agreed to transport us, at $60 million a pop.

Why would nations compete in space, anyway? If such competition occurs, how might it affect us? What should we have in space: Kumbaya or Starship Troopers? Or is the answer somewhere in between?

The "Moon race" of the 1960s was a Cold War exercise of soft power projection, meaning that it involved no real military confrontation, but rather was a competition by nonlethal means to determine which country had superior technology, and by extension, *the* superior political and economic system. In short, it was largely an international propaganda struggle. Simultaneously, the two countries also engaged in a hard power struggle in space to develop ever-better systems to observe and monitor the military assets of the other. There was little public debate associated with this struggle, indeed, much of it was kept secret. As the decade passed, military space systems became increasingly more capable and extensive. Over time, they largely replaced human intelligence assets monitoring our adversaries' strategic capabilities and intentions.

The United States very publicly won the race to the Moon, giving rise to a flurry of pronouncements about everyone's peaceful intentions for outer space, while the larger struggle continued to play out behind the scenes. NASA's replacement effort for the concluded Apollo program, the space shuttle project, promised to lower the costs of space travel by providing a reusable vehicle that would launch like a rocket and land like an airplane.

Because of the need to fit under a tightly constrained budgetary envelope, and for a variety of other technical reasons, the shuttle did not live up to its promise as a low-cost "truck" for space flight. However, the program resulted in a fleet of five operational spacecraft that successfully flew 133 missions over the course of its 30-year history.

Although some in American space circles have called it a policy failure, the shuttle had some interesting characteristics that caused it to be considered a military threat by the USSR. An early shuttle mission had its crew retrieve an orbiting satellite, Solar Max, for repair. Later missions grappled balky satellites and returned them to Earth for refurbishment, repair, and re-launch. This capability culminated with a series of shuttle missions to the Hubble Space Telescope (HST) that conducted on-orbit servicing tasks, ranging from correcting the flawed optics of the original telescope (the first service mission) to the routine upgrading of sensors, the replacement of solar arrays and main computers, and the reboosting of the telescope to a higher orbit. The significance of these missions was that the HST is basically a strategic reconnaissance satellite: It looks up at the heavens rather than down at nuclear missile sites from orbit. The Hubble repair missions documented the value of accessing orbital assets with people and servicing equipment.

Another relatively unnoticed series of shuttle missions demonstrated the value of advanced sensors. As a large, stable platform in orbit (its orbiting mass was almost 100 metric tons), the shuttle was able to fly very heavy, high-power payloads that smaller robotic satellites could not. The Shuttle Imaging Radar (SIR) was a synthetic-aperture instrument that could obtain images of Earth from space by sending out radar pulses as an illuminating beam. It was able to image through cloud cover, day or night, all over the Earth. In a stunning realization, we found that it could also image subsurface features from space—in particular, the SIR-A mission mapped ancient riverbeds buried beneath the sands of the eastern Sahara.[6] The strategic implications of this discovery were immense; as most land-based nuclear missiles are buried in silos, the use of sensors like imaging radar means that they cannot remain hidden.

These new capabilities, provided by the space shuttle, had significant policy implications for the Soviets. To them, it seemed that the shuttle was a great leap forward in military space technology, not the "policy failure" bemoaned by American analysts. With its capabilities for on-orbit satellite servicing and as a platform for advanced sensors, the shuttle became a threat that had to be countered. The USSR responded with *Buran*, its version of a space shuttle, which looked superficially similar to the American version. The *Challenger* accident showed that the shuttle was a highly vulnerable system in many respects; even as the Soviets developed *Buran*, the American military had already decided to withdraw from the shuttle program.

During the 1990s, we saw a revolution in tactical space—the use of, and reliance on, space assets on the modern battlefield. The global positioning system (GPS) has made the transition to the consumer market, but it was originally designed for our troops to instantly know their exact locations. A global network of communications satellites carries both voice and data, and interfaces to the partly space-based Internet, another innovation originally built for military technical research. Now, the entire world is connected and plugged in, and spacebridges are important components of that connection. Fifty years into the Space Age, we are all vitally dependent, both economically and militarily, on our satellite-based assets; space is, by default, in control of Earth's economic sphere. Whoever controls cislunar space controls what happens on Earth.

Most people do not know about the multitude of satellites in various orbits around the Earth that affect their daily lives. We rely on satellites to provide us with instantaneous global communications that affect almost everything we do. We use GPS to find out both where we are and where we are going. Weather stations in orbit monitor the globe, alerting us to coming storms so that their destructive effects can be minimized. Remote space sensors map the land and sea, permitting us to understand the distribution of various properties and how they change with time. Other satellites look outward to the Sun, which controls the Earth's climate, and "space weather," which influences radio propagation. The satellites orbiting the

Earth provide us with phenomenal amounts of information. Fortunately, they are not yet self-aware—but the people who operate them are.

All satellites are vulnerable. Components constantly break down and must be replaced. New technology makes existing facilities obsolete, requiring high-cost replacements. A satellite must fit within and on top of the largest launch vehicle we have. Spacecraft thus have practical size limits, which in turn limits their capabilities and lifetimes. Once a satellite stops working, it is abandoned and a replacement must be designed, launched, and put into its proper orbit, all at great cost.

Although satellite aging is normal and expected, catastrophic loss, either accidental or deliberate, is always a concern. Encounters between objects in space tend to be at very high velocities. The ever-increasing amounts of debris and junk in orbit, such as pieces of old rockets and satellites, can hit functioning satellites and destroy them.[7] North American Aerospace Defense Command (NORAD) carefully tracks the bigger pieces of space junk. Some spacecraft, such as the International Space Station, can be maneuvered out of the path of big chunks of oncoming debris, but smaller pieces, say, the size of a bolt or screw, cannot be tracked and avoided. Such debris could cripple a satellite if it collides with some critical part of the vehicle.

Antisatellite warfare (ASAT) is another possible cause of failure. It has long been recognized that satellites are extremely vulnerable to attack, and both the US and the USSR experimented with ASAT warfare during the Cold War. ASAT takes advantage of the fragility of these spacecraft to render them inoperative. This can be done with remote effectors, such as lasers to "blind" optical sensors. The simplest ASAT weapon is a kinetic impactor. By intercepting a satellite with a projectile at a high relative velocity, the satellite is rapidly and easily disintegrated and rendered worthless.

Despite their vulnerability, the destruction of space assets has seldom happened by accident and never as an overt act of war. They are left alone because they are not easy to get to. Some orbiting spacecraft occupy low Earth orbit (LEO) and are accessible to interceptors, but many valuable strategic assets reside in the much higher orbits of middle Earth orbit

(MEO) between 3,000 and 35,000 kilometers, and in geosynchronous Earth orbit (GEO) at 35,786 kilometers. Such orbits are difficult to reach, requiring long transit times and complex orbital maneuvers, which quickly reveal themselves and their purpose to ground-based tracking.

After a booster failure in 1998, a communications satellite was left in a useless transfer orbit. Engineers at Hughes, the makers of the satellite, devised a clever scheme to send the satellite to GEO using a gravity assist from the Moon. This first "commercial" use of a flight to the Moon saved the expensive satellite for its planned use.[8] One aspect of this rescue is seldom mentioned but it attracted the attention of military space watchers everywhere. This mission dramatically illustrated the importance of what is called "situational awareness" in space. Most trips to GEO travel from LEO upward; this satellite came down from the Moon, approaching GEO from an unobserved (and at least partly unobservable) direction, one not ordinarily monitored by ground-based tracking systems.

With few exceptions, we are not able to access satellites to repair or upgrade them. Satellites must be self-contained. Once they stop working, they are replaced. Sometimes favorable conditions allow us to be clever and rescue an asset that had been written off, but the system is not designed for such operation. The current spaceflight paradigm is a use and throwaway culture. Our history with the space shuttle program demonstrates that this template need not be the way of conducting business in the future. What is missing is the ability to get people and servicing machines out to the various satellites in all their myriad locations. Reaching LEO is easy, but MEO and GEO cannot be accessed with existing space systems. Yet from the experience of the shuttle and the ISS, we know that if these satellites could be visited, a revolution in the way spaceflight is approached might be possible.

A system with the ability to routinely go to and from the lunar surface is also able to access any other point in cislunar space (see figure 6.1). Our next goal in space should be to create the capability to inhabit the Moon and live off its local resources with the goals of self-sufficiency and sustainability, including learning the skills of propellant production and the refueling of cislunar transport vehicles. Eventually, we can export lunar propellant to fueling depots throughout cislunar space. In short, by going to the Moon, we create a new and qualitatively different capability for space access, a "transcontinental railroad" in space. Such a system would completely transform the paradigm of spaceflight. We can develop serviceable satellites, unlike current ones designed to be abandoned once they fail. This new capability will allow us to create extensible, upgradeable systems. The ability to transport people and machines throughout cislunar space permits the construction of distributed instead of self-contained systems. Such space assets are more flexible, more capable, and more readily defended than conventional ones.

With knowledge of these possibilities, questions arise as to how close we are to developing such a system and if such a paradigm shift for spaceflight is desirable. Are we still in a space race, or is that an obsolete concept? Answers to these questions are not at all obvious. We must understand and consider them fully, as this information is known or available to all spacefaring nations to adopt and adapt for their own use.

The previous space race to land a man on the Moon was a demonstration to the USSR and to the world of our technological superiority. The July 1969 landing of Apollo 11, by any reckoning, gave us technical credibility for the Cold War endgame. It was a huge win for United States and a serious blow to communism. Fifteen years after the moon landing, President Reagan advocated the development of a missile defense shield, the so-called Strategic Defense Initiative (SDI). Although many in the West disparaged this as technically unattainable, the Soviets took the program very seriously. Because the United States had already succeeded in completing a very difficult technical task, the manned lunar landing, something that the Soviet Union could not accomplish, they did not question our technical skill or our resolve. The Soviets knew that a deployment of a US missile shield would instantly render their entire nuclear strategic capability useless.

Not only did the Apollo program achieve its literal objective of landing a man on the Moon (propaganda, soft power), but it also achieved its more abstract objective of intimidating our Soviet adversary (technical surprise, hard power).[9] Apollo thus played a significant role in ending the Cold War, one far in excess of what many scholars believe. Similarly, our two follow-on programs of shuttle/station, although fraught with technical issues and deficiencies as tools of exploration, were significant in our understanding and pursuit of human spaceflight, providing us with a way to get people and machines to satellite assets for construction, servicing, extension, and repair. We learned how to assemble very large systems in space from smaller pieces. From our experience constructing the ISS, mastery of these skills suggests that the construction of new, large distributed systems for communications, surveillance, and other tasks is possible. These new space systems would be much more capable and enabling than existing ones.

Warfare in space is not as it is depicted in science-fiction movies, with flying saucers blasting lasers at speeding spaceships. The real threat from space warfare is the denial of assets: Communications satellites are silenced, reconnaissance satellites are blinded, and GPS constellations are made inoperative.[10] Possessing this capability completely disrupts command and control and compels reliance on terrestrially based systems, making force projection and coordination more difficult, cumbersome, and slower.

By testing ASAT weapons in space, China has indicated that it fully understands the military benefits of hard space power.[11] It also has a well-developed lunar program. Currently, China's ambition of flags and footprints on the Moon represents soft power projection. Since only the United States has done this in the past, China would celebrate a successful manned lunar mission as a great propaganda coup. Sending taikonauts, as the Chinese call their astronauts, beyond low Earth orbit is a statement of technical parity with the United States. Historically known for taking the long view, often spanning decades, unlike the short-term view that America favors, China understands and appreciates the strategic importance and value of cislunar space.[12] Thus, although initial Chinese plans for human lunar missions do not feature resource utilization (ISRU), they know from the technical literature that this activity is both possible and enabling.

The Chinese are also aware of the value of the Moon as a "backdoor" to approach other levels of cislunar space, as demonstrated by the rescue of the Hughes communications satellite. The lunar mission Chang'E 2 is an instructive case in point. Ostensibly a global mapper, the Chang'E 2 spacecraft was launched to the Moon in October 2010. It successfully inserted into lunar orbit and spent the next eight months mapping the surface in detail. Then, the mission took a strange turn; after leaving lunar orbit in June 2011, the Chang'E 2 spacecraft slowly traveled to the Sun-Earth L-2 point (fixed in space relative to the Earth) where it proceeded to loiter for the next eight months. Departing the L-2 point in April 2012, the Chang'E 2 spacecraft then intercepted and flew past and within about three kilometers of Toutatis, a near-Earth asteroid orbiting the Sun. The spacecraft successfully sent images and other data of its encounter back to Earth.

This mission profile is significant in terms of space defense. The Chinese demonstrated their ability to dispatch and maneuver a craft throughout cislunar space, including the tasks of rendezvous and interception, and to command and operate this vehicle throughout the multiyear duration of the mission. Loiter, interception, and action on command are three pillars of antisatellite warfare. Moreover, a spacecraft on an interception path from above—rather than below, as would be the case for antisatellite missions launched from Earth—is much more difficult to detect and track. In short, with the Chang'E 2 mission, China demonstrated that it possesses the ability to base ASAT weapons in deep cislunar space and intercept trans-LEO space assets at will, assets that have very little in the way of defensive capabilities.

If space resource extraction and commerce is possible, a significant question emerges: What societal paradigm shall prevail in this new economy? Many New Space advocates assume that free markets and capitalism are the obvious organizing principles of space commerce, but others may not agree. For example, to China, a government-corporatist oligarchy, the benefits of a pluralistic free-market system are not obvious. Western capitalism is successful because of the enforcement of and respect for contract law. Implementation of capitalism in the developing world has met with mixed results, and truly free markets do not exist in China. What will the organizing principle of society in the new commerce of space resources be: the rule of law or authoritarian oligarchy? An American win in this new race for space does not guarantee that free markets will prevail, but an American loss could ensure that free markets would not emerge and drive expansion on this new frontier. The struggle for soft power projection in space is ongoing.

Once it was decided upon in 1961, President John F. Kennedy laid out the reasons why America had to go the Moon.[13] Among the many ideas he articulated, one stands out: "Whatever men shall undertake, free men must fully share." This is a classic expression of American exceptionalism, the idea that we explore new frontiers not to establish an empire, but to ensure that our political and economic system prevails, a system that has created the most freedom and placed the most new wealth in the hands of the greatest number of people in the history of the world. This is a statement of both soft and hard power projection; by leading the world into space, we guarantee that space does not become the private domain of powers who view humanity as cogs in their ideological machine, but rather as individuals to be valued and protected, and given the opportunity and latitude to innovate and prosper.

The Moon is the first destination beyond LEO because it has the material and energy resources needed to create a true spacefaring system. Recent data from the Moon show that it is even richer in resource potential than we had previously thought; both abundant water and near-permanent sunlight are available at selected areas near the poles. We go to the Moon to learn how to extract and use those resources to create a space transportation system that can routinely access all of cislunar space, with both machines and people. Such a system is the logical next step in both space security and space commerce. This goal for NASA makes the agency relevant to important national interests. A return to the Moon for resource utilization contributes to national security and economic interests, as well as scientific ones.

We are in a new space race, and it is a struggle that has both hard and soft power dimensions. This race is real and more vital to our country's future than the original one, if not as widely recognized and appreciated. The hard power aspect is to confront the ability of other nations to deny us access to our vital satellite assets in cislunar space. The soft power aspect is a question: How shall society be organized in space? Both concerns are equally important and both can be addressed by lunar return. Will space remain an ever-shrinking sanctuary for science and public relations stunts, or will it be a true frontier, opened wide to scientists and pilots, as well as miners, technicians, entrepreneurs, and settlers? Decisions made now will decide the fate of spacefaring and affect our national economic and security status for generations.

The Role of Public Opinion in Spaceflight

A familiar refrain about the civil space program is that we must somehow get and keep the American people "excited" about space. NASA has spent a great deal of energy pursuing this elusive goal. Its outreach efforts are designed to convince the taxpayers that spending money on space is a good investment. The most common approach is an appeal made to impress upon the American people how the many benefits from spin-off technologies, goods, and capabilities inspired or created by space research and development have affected their lives in positive ways.

The Aldridge Commission received a presentation from NASA Public Affairs that contained 50 years of polling data on the question, "Do you support the American space program?" The poll numbers on this question have bounced around through the years, ranging from close to 60 percent to as low as around 40 percent. Surprisingly, no matter what the agency was doing, how it was faring, what disasters it endured or the triumphs it had achieved, the typical breakdown was roughly 50–50, plus or minus 10 percentage points. This result, nearly constant over the course of the 50-year history of the space program, is as rock-solid as almost any polling number in existence over a similar time span.[14]

Yet, NASA wrings its hands over this result: "How can we excite the people? If we could just come up with the correct public relations plan, the public and Congress will shower us with money and support!" I believe that these numbers have a different significance. If your poll results are always around 50–50, then, in a fundamental sense, people are indifferent about what you are doing. Apparently, the public really doesn't fixate on what NASA does. True enough, many do have a fascination with spaceflight; attendance at the National Air and Space Museum is consistently the highest of all the museums on the National Mall in Washington. But as with any museum visit, their curiosity is easily satiated, and few dwell on national strategic and economic goals and objectives in space on a daily basis.

While NASA sees its 50–50 polling approval as a problem, I see it as an opportunity. In broad and vague terms, people support our space program. It is a source of national pride, and Americans don't want to see NASA on the chopping block. They like the idea of going to new places and making new discoveries; they just don't center their thinking on the sausage making of space policy. What they want from their government is a space program that does interesting things and not too many dumb ones, with programs that inspire the country and make us smarter, hopeful, and proud.

Given this relatively benign public mood and a funding level almost literally "in the noise" compared with other federal programs—at less than 0.3 percent of the federal budget, *much* smaller than most believe it to be—NASA's strategic direction should focus on the incremental buildup of our capability to go farther, stay longer, and develop and increase human "reach" beyond low Earth orbit, first, into cislunar and then into interplanetary space. Our Moon is situated where it can play an important role in this buildup, since it is the first place beyond low Earth orbit with the resources needed to develop and expand our spacefaring capability. Initially, this means oxygen and hydrogen—vital, consumable resources necessary to support a human presence, and as rocket propellant for refueling spacecraft. Provisioning in space begins here.

Perhaps the public doesn't care about the Moon or even the space program in general. But even if this is true, it is irrelevant. Few concern themselves with the requirements and properties of infrastructure development such as railroads or highways, yet no one denies their value, nor does it stop their productive use of them. As a modern, technical society, we depend upon space and the assets and resources found there, for a wide variety of purposes. In order to take advantage of these opportunities, we need freedom of movement on the ocean of space—the ability to go where we need, whenever we want to. The development of lunar resources holds great promise by giving us the flexibility to pursue a set of long-term goals in space—goals that will ultimately allow us to go anywhere, for any amount of time, to do almost any job we can imagine, as well as doing those things that we can not yet imagine.

Moonrush: Issues in Private Sector Lunar Activities

A number of American companies, at differing levels of involvement and with greatly differing degrees of technical credibility, claim to be attempting lunar spaceflight. A stimulus to this activity is the Google Lunar X-Prize (GLEX), a $20 million contest to safely land a payload on the Moon and conduct a number of specified milestone activities.[15] Although this seems like a stunt, the rationale behind GLEX is serious. Prizes are employed by other technical fields of endeavor to stimulate development and innovation. Winning a prize has multiple benefits: It awards money, confers prestige by succeeding over competitors, creates acclaim, and generates business opportunities. Competing is also a good way to compress timescales of technical innovation and development: to win the prize, achieve "x" by "y" time.

Although space entrepreneurs and experts often tout the value of prizes in stimulating new technical accomplishment, their efficacy in the field of space to date has been less than impressive. The Annsari X-Prize for the first commercial suborbital flight was won in 2004, but as of 2015, no other commercial suborbital flight has taken place.[16] Space businessman Robert Bigelow established America's Space Prize, a $50 million award for the first commercial provider of the transport and return of five human passengers to LEO. The prize was announced on December 17, 2003, the hundredth anniversary of the first Wright Brothers flight, and it expired in January

2010 without a single attempt to claim it. The GLEX was announced in 2007 and had a deadline of 2012, a deadline that has been extended twice, first to 2014 and then to the end of March 2015. A third extension to the end of 2017 was recently announced. I do not inventory these dismal statistics to disparage prize offers. I merely point out that they have a poor record of creating new capabilities.

Most discussions about lunar resources focus almost exclusively on the technical issues associated with extraction, transportation, and use. Little has been offered on the legal issues involved in lunar or extraterrestrial mining—staking a claim, in other words, just as a miner does on Earth. This vacuum exists for a very straightforward reason: No one knows the legal status of commercial space mining and planetary surface activity.

Several international treaties, the most pertinent of which is the 1967 U.N. Outer Space Treaty, set the current legal regime for space activities.[17] Signed by 129 countries, including all of the major spacefaring nations, the treaty bans nuclear weapons in space and prohibits any nation from establishing territorial claims on extraterrestrial bodies. This formulation left open the question of private development and ownership, although the treaty states, "Outer space, including the Moon and other celestial bodies, shall be free for exploration and use by all States without discrimination of any kind, on a basis of equality and in accordance with international law, and there shall be free access to all areas of celestial bodies." Note well the language: "free for exploration and use by all States." That wording would appear to guarantee the rights of a nation to mine the Moon, extract a product, and then—what?

Certainly one would suppose that this language ensures that a government facility could manufacture rocket propellant to use in its own vehicles. But does it permit a private company based in that nation to make the same product and then offer it for sale on the open market? Certainly the Federal Aviation Administration (FAA) can issue restrictions on American companies in regard to impinging upon the activities of another American company—say, for example, Moon Express landing a vehicle near an installation of Bigelow Aerospace inflatable habitats on the

Moon. But who else is obliged to observe those restrictions? International companies that launch from their own soil do not require FAA commercial licenses. Unless some reciprocal agreement is reached between all of these nations, their private companies do not have to respect the access and "control zone" rights of other nations' companies.

The situation becomes even murkier when considering the possible interactions of a private American company on the Moon and the national representatives of a foreign power. Suppose another country such as China decided for whatever reason to land a government-funded, military-controlled spacecraft on a patch of lunar territory that the FAA had previously set aside for the exclusive use of Bigelow Aerospace. Legally, the FAA license has nothing to do with China, which is not bound to observe any restrictions. When international relations are peaceful and productive, conflicts are unlikely to arise. But political situations change, sometimes at the drop of a hat, and certainly on timescales shorter than industrial development cycles.

Prime locations on the Moon, as on any other extraterrestrial object, are not limitless, and access to and use of the most desirable and valuable sites for resource prospecting and harvesting may become contentious. In terms of water production for rocket fuel and life support consumables, ideal sites are in zones of enhanced duration sunlight ("quasi-permanently lit areas") near the Moon's poles, proximate to permanently shadowed regions and thus deposits of water ice. At such locales, electrical power can be continuously generated in order to extract the nearby water ice. There may be only a few dozen zones where initial ice harvesting facilities may be operated with reasonable efficiency, on which more prospecting data will give us a better picture. If this turns out to be the case, then who gets the rights to produce the product? What constitutes staking a claim? First come, first serve? Or does might make right?

This issue leads us to consider the presence and role of the federal government in space. I contend that a strong federal presence in space is necessary to ensure that our rights are established and that our values are protected and promoted. In the hypothetical context mentioned of Bigelow and

China mentioned before, a single American company facing a determined nation-state is not likely to prevail in a manner favorable to the interests of free market capitalism. Legal recourse on Earth would be limited—more likely nonexistent. It is also unlikely that the United States would go to war over the infringement of some corporate plot of land on the Moon, at least during the early stages of commercial space. However, when the federal government establishes a presence, it serves notice to the world that we have national interests there. Their presence makes any infringement on the property and access rights of American corporations less likely to occur in the first place—and more easily resolved if such a situation arose, creating a much more favorable climate for private investment in space activities.

There is no reason to assume that all nations will voluntarily cooperate in space, if for no other reason than nations do not behave this way on Earth. Sometimes national rights of way and access to resources must be guaranteed by a physical presence, backed up with threat of force. This is the way of life at sea here on Earth and the reason we have a blue-water navy—not only to defend our country but also to project power and protect our national interests abroad. Historically, the navy has conducted exploration and goodwill tours in peacetime and power projection in times of tension and war. A space navy could do likewise as humanity moves outward into the solar system.

Ultimately, we will need to face up to our national and collective responsibilities to protect American commerce and interests wherever they reside. Given the cost risk of opening up space to commerce, companies need assurance that government can, and will, help protect their investment. In the very near future, our theater of operations will include cislunar space. The idea that the private sector alone can develop near Earth space is not realistic, nor even advisable. It remains a dangerous, unpredictable world, and clear-thinking leaders need to plan for future confrontations, if only, so that they can be avoided. Any display of weakness will be exploited—and not to our benefit.

9

A Visit to the Future Moon

I believe the central, near-term goal for the American civil space program needs to be a permanent return to the Moon. Once we do this, we create a new and versatile spacefaring infrastructure, one capable of extending human reach beyond low Earth orbit (LEO) into the solar system. If we eventually head in such a direction, what might we expect to see in the future? What benefits will accrue from this direction and how might they develop over time? Here I envision a future for humanity on the Moon and a series of steps and events that are most likely to occur, along with their appropriate implications. On the Moon, we will begin to use and settle space for a variety of beneficial purposes.

Early Activities

In one sense, our lunar return has already begun through the series of robotic spacecraft that have mapped, measured, and surveyed the Moon over the past decade. Most of these missions were orbiters, loaded with a variety of sensors and instruments designed to measure physical properties in almost every part of the electromagnetic spectrum. These data, converted into maps showing shape, size, composition, and physical state, have given us a clearer picture of the makeup and evolution of our nearest neighbor. The Moon is probably the best-mapped object in the solar system, while parts of Earth's ocean floor are more poorly known than the lunar far side. These survey maps allow us to evaluate the Moon as a planetary object. The regional inventory of its resources determined from orbit show that the Moon possesses what we need to create a new spacefaring capability. The Lunar Reconnaissance Orbiter (LRO) continues to give us the knowledge that fuels research and produces new discoveries.

Impactors and landers have also added critical detailed information for small, selected areas on the Moon. One of the most important pieces of information came from the LCROSS impactor. In this mission, the upper stage of the LRO launch vehicle was crashed into one of the cold, dark regions of the lunar south pole. The collision was observed by a small satellite that had followed the impacting upper stage and by the LRO spacecraft, already in lunar orbit. The ejected material from this impact conclusively demonstrated that water ice is present within this cold trap. The amount is estimated at about 7 weight percent at this location. The ejecta plume also threw up other volatile species, including ammonia (NH_4), methane (CH_3), carbon monoxide (CO), and some simple organic molecules.[1] These data suggest that the volatiles of the Moon's polar regions are likely of cometary derivation. With this information, we can state with a strong degree of confidence that the materials needed for permanent human habitation on the Moon are present in the Moon's polar regions.

We have located and quantified the areas near the Moon's poles that receive the most sunlight over the course of a year (figure 3.1). These lit regions are close to deposits of water ice and other volatiles. Maps produced by LRO and its orbital companions will be crucial in locating likely sites for resource processing on the Moon. In addition to the direct sampling of lunar ice by LCROSS, several different remote measurements support the presence of significant amounts of polar water. The M^3 spectral mapper on India's Chandrayaan-1 spacecraft found evidence for hydroxyl molecules (OH) at high latitudes (figure 9.1), which migrate poleward to possibly serve as a source for polar water.[2] A small impactor from Chandrayaan-1 (the Moon Impact Probe) found a tenuous water vapor cloud over the south pole (probably water molecules en route to their ultimate site of

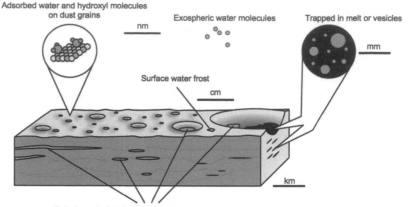

Figure 9.1. Schematic showing the five modes of occurrence of lunar water. Water is found within melts formed deep inside the Moon and sampled by volcanic glasses and minerals. Exospheric water occurs as rare molecules that bounce around the Moon in the space just above the surface. Adsorbed water is found as a monolayer of molecules on dust grains; these molecules increase in abundance with increasing latitude (decreasing mean surface temperature). Surface water frost of more substantial quantity is seen within the dark, cold areas near the poles. Larger amounts of water ice may occur near the pole at shallow depths (a few meters or less) in substantial amounts (millions of tons).

deposition in a polar cold trap). Mini-RF radar images show high diffuse backscatter in some polar craters (see figure 5.1). Ultraviolet spectra and laser reflections indicate the existence of water frost on the surface of the floors of some polar craters. Neutron measurements over the poles indicate extensive amounts of hydrogen. These data support our understanding about the presence of significant amounts of ice at both poles, as much as 10 billion tons at each pole.

Despite the abundant new data, in order to achieve a permanent lunar presence, we must understand and map the variations in polar water content on meter-scales, laterally and vertically. The physical properties of the ice must be determined to plan for excavation and water extraction. We must find areas of the highest water concentration that are closest to the areas of "quasi-permanent" sunlight, so as to make future water processing most efficient. These properties and others can be obtained from additional robotic surface exploration. The ideal way to get the highest quality data is to land a nuclear-powered surface rover, similar to the current Mars Science Laboratory, and conduct an extended traverse across the polar region to find and map out the best areas.[3] Identically equipped rovers should be sent to each pole; although we suspect that both north and south poles possess significant volatile deposits, the scouting of both areas by two rovers would help us be certain that we locate the outpost near the highest grade deposits.

Short of this fairly sophisticated level of exploration, a series of smaller missions could gather preliminary information on polar volatiles. One example of an inexpensive mission is to fly a pallet of about a dozen small impactors (hard landers) that would be individually deployed and landed to gather surface compositional and physical data from multiple points. Although less desirable than the detailed, continuous information that a properly equipped rover would provide, this approach may be a good strategy to collect widespread, detailed data in a short period of time for a small amount of money.

When enough prospecting data have been obtained, the next most pressing need is to demonstrate the process of resource extraction and storage on the Moon. Although water extraction is probably the simplest processing of extraterrestrial materials imaginable, in order to be taken seriously by some in the space engineering community, an actual end-to-end system demonstration is needed. Such a demo mission could be quite small; a fixed lander in the sunlight, fed with feedstock from the shadowed area, could heat the soil, collect the water vapor, liquefy it, and store it. Once this demonstration has been accomplished, the production of large amounts of water becomes merely a matter of scale.

Some lingering mysteries about the lunar surface environment also need to be addressed. It has been postulated that the passage of the day/night line (the terminator) across the surface induces an electrical charge, one of possibly dangerous magnitude. This effect could be measured and that possible risk retired through a series of measurements from a fixed lander over the course of a lunar day. Observing the postulated levitation of fine-scale dust by electrical fields should also be studied on the surface, although evidence obtained recently from the orbital LADEE mission suggests that this phenomenon, if it occurs at all, is minor and of local extent.[4]

Consolidating Our Lunar Presence

As previously described, I believe that the most efficient and least expensive way to return to the Moon will require performing much of the preliminary, early work with robotic assets, followed later by people.[5] In the early stages of lunar return, robotic machines operated from Earth can begin the harvesting and processing of lunar water. We should initially plan to build up enough capability to fuel a return trip back to Earth before humans arrive. Such a capability requires the production of about 100 tons of water per year. This isn't as great a quantity as one might imagine: 100 tons of water is roughly the amount contained in a tank the shape of a cube 15 feet (4.5 m) on a side, or roughly the volume of water in a single backyard swimming pool. Because water is the most enabling resource with the widest possible range of use, it is the first priority for utilization.

Significant mining activity on the Moon will require power and lots of it. Fortunately, there is enough surface area in the polar quasi-permanent sunlit zones (see figure 3.1) to establish networks of multiple solar array power stations. A single station could consist of a tall (~10–20 m), narrow (~2–3 m wide) array of solar cells that can be articulated around its vertical axis (see figure 7.1) to track the Sun as it slowly moves around the horizon over the course of a lunar day. Such an individual station would be low mass (~1 ton). As a modular system, these pieces could be connected together to provide whatever level of power is needed. Initial robotic mining capability would require roughly 150 kilowatts, power that could be provided by eight to ten individual power stations. As outpost capability and size grows over time, additional power stations delivered from Earth can satisfy generating needs. This potential for growth in a surface power system is possible up to about the ten-megawatt level, after which we would probably need to consider the deployment of a nuclear reactor. A thorium molten salt reactor could be sized to provide virtually unlimited power (hundreds of megawatts) for a wide variety of uses and while initially supplied entirely from Earth, could ultimately be operated from locally mined sources of thorium on the Moon.

In civil engineering, one of the most important material resources on Earth is "construction aggregate"—the sand, gravel, and cement building materials that make up the infrastructure of modern industrial life. Aggregate is easily one of the most important and valuable economic resources of all mined terrestrial materials, more so than gold, diamonds, or platinum. We depend on aggregate for many different types of objects; they are the fundamental building materials of roads and structures. The use of aggregate in building goes back to ancient civilizations, such as the concrete used for construction in ancient Egypt. The Romans devised a recipe for a concrete so durable that the molded arches, walls, and self-supporting dome of the Pantheon, built more than two thousand years ago, still stand today. Aggregates in terrestrial use typically depend on a lime-based cement that bonds the particulate material together. Both lime (CaO) and abundant water are needed to make concrete on Earth.

By necessity, a permanent presence on the Moon will require an infrastructure that uses as much local material as possible. Aggregate materials probably will become the primary building blocks of industrial society off planet, just as it has on the Earth. The composition and conditions of local materials will require some adjustments as to how we use lunar aggregate. A quick assessment reveals some interesting parallels, as well as differences, with terrestrial use.

On Earth, gravel pits are carefully located to take advantage of the sorting and layering produced by natural fluvial activity. We harvest gravels from alluvial plains and old riverbeds, where running water has concentrated rocks, sand, and silt into deposits that can be easily excavated, loaded, and transported to sites of construction. The highly variable currents, as well as the velocities of flow of our terrestrial streams and rivers, sort the aggregate by size. This natural sorting creates layers of gravel- to cobble-sized stones for the fastest flowing waters. Finer-grained material is likewise concentrated where water speeds are low, and sand and silt settles out from the suspended sediment (the "bed load").

Unlike the aggregate processing by water on Earth, lunar surface rock has already been disaggregated into a chaotic upper surface layer (regolith) by impact. Regolith is ground-up bedrock; impacting objects of all sizes constantly pummel the surface, breaking, fracturing, and grinding up the Moon's bedrock, a process of impact that has greatly slowed from the much higher level experienced earlier in lunar history. The regolith is a readily available building material for construction on the lunar surface. It is an aggregate in the same sense as on Earth, but with some significant differences. We could make lime and water from the Moon's surface materials but that would require too much time and energy. Thus, we should adapt and modify terrestrial practice to take advantage of the unique nature of lunar materials. The fractal grain size in the regolith means that we can obtain any specific size fraction we want through simple mechanical sorting (raking and sieving). Instead of water-set, lime-based cement, we can use the glass in the regolith to cement particulate material together, that is, sinter the aggregate into bricks and blocks, as well as roads and landing pads, using thermal energy (figure 9.2). Both passive solar thermal power

Figure 9.2. Robotic rover carrying microwave sintering equipment to fuse local regolith into ceramic pavement for use as a landing pad for spacecraft. Power for melting the soil is provided by solar array at center.

(concentrated by focusing mirrors) or electrically generated microwaves can provide the energy to melt grain edges into a hard, durable ceramic.

The use of aggregate on the Moon will likely be gradual and incremental. Our initial presence on the Moon will be supported almost entirely by materials and supplies brought from Earth. As we gain experience in using lunar resources, we can incorporate local materials into the structures. Simple, unmodified bulk soil is an early useful product. It can be used in building berms to protect an outpost from the rocket blast of arriving or departing spacecraft, and to cover surface assets for thermal and radiation protection. The next phase will be to pave roads and launch/landing pads to limit the amount of randomly thrown dust and to provide good traction for a multitude of wheeled vehicles supporting the outpost. The fabrication of bricks from regolith will allow us to construct large buildings, initially consisting of open, unpressurized workspaces and garages, but ultimately habitats and laboratories. The new technology of three-dimensional (3D) printing will allow nearly autonomous machines to construct the lunar outpost through the use of regolith aggregate assembled into structures by 3-D printers working in conjunction with Earth-controlled construction robots.[6] Making glass by melting regolith can produce building materials of extreme strength and durability; anhydrous glass made from lunar soil is stronger than alloy steel, with a fraction of its mass.

Metals are abundant in the Moon and can be extracted from the local materials. The basic process is one of simple chemical reduction, accomplished through a variety of low-tech processes, all of which were known to eighteenth-century industry. Carbothermal reduction of ilmenite, an iron and titanium oxide, has been demonstrated in the laboratory to produce oxygen; it also produces native metal as a by-product. The use of fluorine gas as a reducing agent has also been well studied. Metal production techniques require large amounts of electrical power, as it takes significant energy to break the tight metal-oxygen bonds in common rock-forming mineral structures. For this reason, it is likely that metal production will come late in lunar industrialization; initial surface structures and base infrastructure pieces are likely to be made from lower-energy products, like composites and aggregate.

Although most products made on the Moon will be used locally, eventually we can export lunar products into space. The gravity well of the Moon is a drawback for large mass delivery—its escape velocity is about 2.38 kilometers per second, much smaller than that of Earth (11.2 km/s), but still substantial. In order to use large quantities of lunar materials for space construction, we need to develop an inexpensive means to get material off its surface. Fortunately, the Moon's small size and lack of atmosphere make this possible by building a system that literally throws material off the Moon into space. A "mass driver" can launch objects off the lunar surface by accelerating them along a rail track using electromagnetic coils that hurl encapsulated material into space at specific velocities and directions.[7] We can collect such thrown material at a convenient location, such as one of the libration points. From there, it is a relatively simple matter to send the material to wherever it is needed in cislunar space. A mass driver is not a science-fiction concept; such systems are used to launch planes from the flight decks of aircraft carriers.[8]

Surface Activities and Exploration

An early goal for lunar return is to become self-sufficient in the shortest amount of time possible. This does not mean that a significant amount of surface exploration and science is not also attainable. By virtue of being on the Moon for extended periods, we will have many opportunities to study lunar processes and history in unprecedented detail. Our scientific tasks include understanding the nature and details of the regolith and its interaction with the space environment (a topic addressable any place on the Moon). Such study has both practical relevance, to better conduct resource processing and to improve product yield, and academic interest, since details of regolith dynamics remain elusive. An example of a simple, easy-to-complete experiment is to dig a trench in the regolith several meters deep.[9] In the exposed wall of this trench would be several billion years of solar and impact history available for our inspection, sampling, and detailed study.

The Moon has experienced the processes of impact, differentiation, volcanism, and tectonism. These processes occur on all rocky planets in the solar system. The antiquity of the lunar surface ensures that near-complete examples of these processes are on display for our enlightenment. By using the Moon as a window into early planetary history, we improve our understanding not only of its history and evolution but also the history and evolution of all the planets. As one example, the Earth and Moon have occupied the same volume of space for the last 4.5 billion years, a space where the impact flux affects both objects. As a result of Earth's highly dynamic surface environment, these ancient events have not been preserved. However, the lunar surface preserves the impact record of the Earth-Moon system dating back to at least 3.8 billion years ago. The study of a multitude of lunar craters can tell us about changes in the impact rates over time, a topic relevant to the extinction and evolution of life in the geologic past.

A common article of faith in many academic and space circles is that robotic spaceflight is the preferred method of scientific exploration. Many famous space scientists, including James Van Allen and Carl Sagan, argued for the superiority of unmanned missions over human ones. Indeed, many phenomena in space, such as plasmas and magnetic fields, cannot be sensed directly by humans, and in some cases, such as detecting the tenuous lunar "atmosphere," the presence of people interferes with the property being measured. I agree that while some scientific activities cannot or should not be done by people, in other areas, a human presence is not just beneficial—it is critical.

The Moon is a natural laboratory, a place where important scientific questions can be answered. The conceptual visualization of the four-dimensional—three spatial dimensions plus time—makeup of planetary crusts is achieved through fieldwork. Fieldwork is not merely a matter of picking up rocks or taking pictures. The "field" is the world in its natural state, where the phenomena we study are on display and where we observe facts and clues that permit us to reconstruct past processes and histories.

A good example of the difference in capabilities between humans and robots is illustrated by the experience with the Mars Exploration Rovers (2003–present). Over the course of their first five years on Mars, these machines traversed many kilometers of terrain, examined and analyzed rock and soil samples, and mapped the local surface. These robotic rovers, giving us an unprecedented view of the martian surface and its geology, have returned many gigabytes of data. They are truly marvels of modern engineering. Yet, after all this extended, robotic exploration, we are unable to draw a simple geological cross section through either of the two MER landing sites. We do not know the origin of the bedded sediments, strikingly shown in the surface panoramas; we do not know whether they are of water-lain sedimentary, impact, or igneous origins. We do not know the mineral composition of rocks for which we have chemical analyses. Without this information, the planet's processes and origins cannot be determined.

Even after more than a decade of Mars surface exploration, we still do not know things about the field site that, given an afternoon's reconnaissance, a human geologist could have deduced. In contrast, we have an incredibly detailed conceptual model, albeit incomplete, of the geology and structure of each of the human visited Apollo landing sites. The longest stay on the Moon for these missions was three days, most of which was spent inside the Lunar Module.

A robotic rover can be designed to collect samples, but it cannot be designed to collect the *correct, relevant* samples. Fieldwork involves the posing and answering of conceptual questions in real time, where emerging models and ideas can be tested in the field. It is a complex and iterative process; geologists can spend years at certain field sites on the Earth, asking and answering different and ever more detailed scientific questions. Our objective in the geological exploration of the Moon is knowledge and understanding. A rock is just a rock—a piece of data. It is not knowledge. Robots collect data, not knowledge.

Because people control planetary exploration robots remotely, it has been argued that human intelligence already guides the robot explorer.

Having done both types of field exploration on Earth, I contend that remote, teleoperated robotic exploration is no substitute for being there. All robotic systems have critical limitations—important sensory aspects, such as resolution, depth of field, and peripheral vision. Robots have even greater limitations in physical manipulation. Picking a sample, removing some secondary overcoating, and examining a fresh surface is an important aspect of work in the field. The physical limitations of teleoperated robots are acceptable in repetitive, largely mechanical work, such as road construction or mining, but in creative, intellectual exploration they are woefully inadequate. The makers of the MER rovers recognized this need by including an abrasion tool to create fresh surfaces; regrettably, it became worn down and unusable after a short period of operation.

Ultimately, we need both people and machines, each with their own appropriate skill bases and limits, to explore the Moon and other planets. Machines can gather early reconnaissance data, make preliminary measurements, and do repetitive or exhaustive manual work. Only people can *think*. And thinking, and then taking action and working with those informed results in real time, is what fieldwork is all about.

On the Moon, we will learn more about the universe, and by doing so, we will also learn *how* to study the universe. Recognizing that people and robots bring unique and only partly overlapping capabilities to the task of exploration, we may find that a combined, specialized approach that builds upon the strengths of both, and that mutually supports the weaknesses in each, is the most efficient and beneficial way to explore.[10] It is easy to conduct thought experiments in how telepresent robots could replace people on planetary surfaces, but we have no real experience in using them. By experimenting with these techniques on the Moon, we can learn the optimum approach for specific exploration tasks. Simple reconnaissance may be conducted with minimal human interaction, but detailed field study might require continuous, real-time human presence. Knowledge of the problems appropriate for each technique is something that can be acquired and understood on the Moon. Such understanding is vital to future exploration and for comprehending other planetary objects.

Building a Transportation Infrastructure

In contrast to the "build, launch, use, throw away, then repeat" paradigm of the past, we seek to create a permanent spacefaring infrastructure that incorporates reusability for as many assets as possible. Although much of the current focus in space development is on reusable launch vehicles, reusability is actually much easier to achieve for vehicles that are permanently based in space. These spacecraft do not have to undergo the thermal and mechanical stresses of launch and reentry. Cislunar transportation consists of multiple steps, including the marshaling of assets at certain points, such as rendezvous and preparation in LEO and L-points (see figure 9.3) followed by transport to the next marshaling area (involving a rocket burn to increase or decrease orbital energy). Since these activities put little stress on vehicle systems, there is no technical reason not to design as much reusability into them as possible.

In the lunar architecture described previously, I call for the building of a 30-ton-class reusable lunar lander (see figure 7.1). The purpose of this

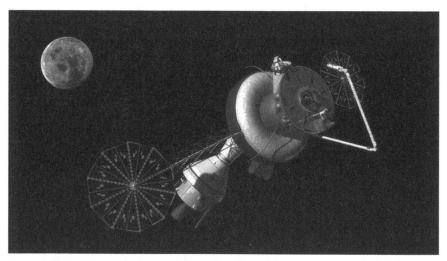

Figure 9.3. Deep space staging node located at Earth-Moon L-1, about 60,000 kilometers above the center of the lunar near side. A staging node can serve as the jumping-off point for missions to the Moon and planets; it contains a habitat for temporary layovers and a propellant depot for the refueling of spacecraft. Future Orion spacecraft shown docked to the transport node.

vehicle is to transport people to and from lunar orbit. By eliminating the need for extended life support, we can make this vehicle smaller than the proposed Altair lander called for by the Constellation architecture. Here, the issue of reusability largely revolves around engine performance and maintenance. A throttleable version of the venerable RL-10 cryogenic engine, used today in the Centaur upper stage, can perform multiple restarts and is a good engine on which to base the creation of a reusable lander. At some point, we will have to change out engines on the reusable vehicles, but they can be made part of a modular system serviceable by suited astronauts and teleoperated machines on the lunar surface. A reusable lander would spend about half of its time on the Moon and the other half in space, at the appropriate staging node, either in low lunar orbit or at one of the L-points. It would be designed to reach its space node with half of its fuel remaining. This permits the lander to make the next descent and landing and then refuel on the Moon with propellant made from lunar water.

Passive, space-based assets require much less work to maintain. Staging nodes, where vehicles can meet and interact to transfer crew, collect cargo, refuel, and the like, will become part of the transportation system. These nodes are actually miniature space stations, complete with their own power, thermal, and attitude control systems. They are much less complex than the ISS, in that they are designed only for a single specialized use and are unoccupied most of the time. However, some maintenance will be needed to keep the nodes functioning correctly. This may include refueling the attitude thrusters and maintaining the electrical and thermal control systems. Transport nodes can be based in several localities, including LEO, an L-point, and in low lunar orbit. The nodes may or may not be associated with a fuel depot.

An orbiting fuel depot is a new technology that can increase the size of payloads placed on the Moon and throughout cislunar space. However, we have much to learn about their construction and operation. The biggest difficulty is learning how to deal with the "boil-off" of extremely low temperature cryogenic liquids. Liquid oxygen boils at −183°C, and liquid hydrogen boils at −253°C. Although we can shield storage tanks from direct exposure to solar illumination with screens, passive thermal radiation from the depot itself will heat these liquids enough to cause evaporation. This problem must be solved to create a permanent space transportation system. Although we do not know how to mitigate this issue at the moment, the solution will probably involve capturing the boil-off gases and recondensing them into liquid.

One way to minimize loss from boil-off is to keep the propellant in a more stable form until it is actually needed. We can transport propellant throughout cislunar space in the form of water, a substance that is easily stored and transferred, and then cracked into the cryogens just before a spacecraft is scheduled to arrive. This would require that the fuel depots of the future also contain a propellant processing system. Such a system would include large solar panels, cryogenic plants, and storage facilities. With this capability, the fuel depot becomes a more complex space station, but it also decentralizes operations throughout the entirety of cislunar space. Note that we will still need these cryogenic processing facilities on the lunar surface in order to refuel arriving spacecraft, along with their obvious importance to the crews who live there.

A complete cislunar transportation system consists of an Earth to LEO transport, multiple staging nodes, fuel depots, transit spacecraft, landers, and the lunar outpost. Such a system permits routine access to the Moon and to all other locations within and throughout cislunar space. For the first time, we will be able to move people and cargo where they are needed, anywhere in cislunar space. Currently, communications satellites at GEO are inaccessible for visits by people. With the new system described here, we can travel to GEO to repair, maintain, or even build new distributed satellite systems of unprecedented power and capability. A communications satellite the size of the ISS could provide uninterrupted communications coverage over a hemisphere, rendering the entire terrestrial cell phone network obsolete in an instant. It could provide enough bandwidth to accommodate thousands of channels of high-definition video, Internet traffic, and personal messaging. Three such complexes would link the entire world with these capabilities; it would generate new wealth and provide endless possibilities for innovation and technology development. In addition, we will be better able to protect sensitive surveillance equipment and other strategic assets. Such capability will make the world safer, in that we would not be rendered blind in the event of aggression and we could better respond to crises, both natural and man-made, that may develop on Earth. The upgrading and enhancement of scientific sensors would also be possible, including such difficult tasks as the servicing of the soon to be launched James Webb Space Telescope, to be located at the Sun-Earth L-2 point and inaccessible to servicing spacecraft with existing systems.

I began this academic journey by explaining how we can use lunar material and energy resources to create a new spacefaring capability—the creation of a permanent transportation infrastructure in space. Such a capability can satisfy all of our requirements to maintain and enhance service satellites, and to open up the Moon (and indeed, the entire solar system) for exploration and development. The rest of the journey—the one that *you* may envision—is now possible.

Exports from the Moon

Until now, I have mainly focused on the development of lunar resources to obtain a foothold on the Moon, but is there anything on the Moon that has economic value *elsewhere,* other than at a lunar outpost? What lunar exports might become profitable in the future and how might such markets, be they private or government, be developed? Is there a "killer app" in lunar resources, a product or service that can create new wealth and actually give us a return on our investment in spaceflight and infrastructure? Many people and nations are keenly aware of the possibilities to realize a profit, and are considering ways to exploit an advantage.

The most obvious lunar product of economic value is water. As previously described, water is an extremely useful substance in space: It can support human life, it serves as a medium for energy storage, and it can be used to make rocket propellant. Thus, for spacefaring nations and companies, by having the ability to purchase useable water already in space, it negates the requirement for them to bring water along from Earth. This option makes their space missions more productive, more routine, and more profitable. A space-based market for water will probably emerge first. Special importance will be given to the availability of propellant at the orbital fuel depots. A good policy would be to husband any surplus water at fuel depot-transport nodes for sale or barter with other spacefaring nations. Such fuel sales could be used to support the flights of other countries on their cislunar missions. It will also find use as fuel for attitude control-orbital maintenance thrusters. At the moment, such thrusters use storable propellant, but if a space-based source for cryogens became available, the satellite builders of the world would soon modify their systems to enable its use.

The idea of generating electrical power in space for transmission back to Earth to be sold commercially has been a staple of lunar development schemes for some time. The Solar Power Satellite (SPS) concept has always faced a major stumbling block; the high cost of launch from Earth of the massive solar arrays make it financially infeasible.[11] A permanent presence on the Moon changes that picture. Solar arrays can be manufactured from lunar surface materials and launched into cislunar space at lower cost, due to the lower gravity of the Moon.[12] In fact, it is likely that if financially viable SPS systems ever become available, they will be made possible only through the use of lunar resources.

An extreme variant of this idea proposes to make the solar arrays in place on the Moon. A small rover rolls along the ground, fabricating amorphous solar cells that are connected and wired together as the rover slowly moves across the lunar surface, manufacturing a solar array that can be tens to hundreds of square kilometers in extent. In the equatorial zones of the Moon, gigantic solar panels farms, with enormous gigawatt-level power output, can transmit to space or directly to Earth via lasers or microwaves. Receivers in either location can collect this power and offer it at commercially competitive rates. To receive constant solar illumination, this system would require the construction of two solar array farms on the equator on opposite sides of the Moon. Seemingly something from science fiction, if undertaken at the appropriate scale, such energy production on the Moon (which has been analyzed economically) is workable.[13]

The possibility of extracting helium-3 from the lunar soil to power fusion reactors on Earth for commercial power generation may be possible within the next few decades, once a determination is made whether such a plan is technically viable or not.[14] If so, helium-3 mining could be a competitor to large-scale solar power generation on the Moon. It would require a significant amount of surface infrastructure to produce commercially useful quantities of the fuel. One wild card in the helium-3 story is that we do not know how much of it might be contained in the polar cold trap volatiles. If these volatile substances are of cometary origin—and analysis of the LCROSS data suggests that they are—helium-3 might be present at roughly solar abundance.[15] Thus, it could be easier and less costly to extract large amounts of helium-3 from polar ice, than from equatorial mare regolith. This is a missing piece of information that will be answered once we are able to send a properly instrumented rover into the polar dark areas on the Moon.

Other lunar products may eventually become economically attractive. We are not imaginative enough to envision them all. The earliest product to have monetary value from export comes from the first product that we make on the Moon—water, in all of its forms. To move through space requires the expenditure of energy in the form of rocket firings. Thus, the freedom of space is energy change. Energy change is a rocket firing. Rocket firings require propellant. To make propellant, we need water. And water is available in large quantity from the polar cold traps of the Moon. Thus, water is the currency of spaceflight. By establishing a resource processing facility on the Moon, we position ourselves to participate in the world markets of the future.

Learning to Live and Work on Another World

Several skills must be mastered and many different technologies must be developed if humanity is to become a multiplanetary species. One recommendation of the 2009 Augustine Committee was to table the notion of selecting destinations in space such as the Moon or Mars and instead work on developing the technology to go anywhere.[16] Then, when we have the technology necessary, we ramp up and go to the planets. This approach, called the "Flexible Path," was quickly embraced by the administration that chartered the committee. Adoption of the Flexible Path was an attempt to distract national attention from the fact that our civil space program was going nowhere.

The largest and most comprehensive expansion of space technology in history was the product of the Apollo program, the antithesis of a "no-destination" effort. The truth is, we get more technology development as a result of the *need* to solve specific problems, problems that arise when we try to do something or go to someplace in space. Confronted with specific issues and needs, technical solutions must be developed or we go nowhere, learn little, solve nothing, and become vulnerable. Historically, a pressing need for answers drives innovation much more quickly and efficiently, than does tinkering around in a hobby shop.

We go to the Moon to learn how to use what it has to offer. One of those offerings is its virtue as a world on which to live and work. Humans have almost no experience with this. The Apollo missions fifty years ago allowed a few people to experience the Moon for a few tens of hours each. From that experience came a dream that has never faded: that a great adventure and future awaits the first people who attempt to make life in space an extended experience. The Moon is our first step. The struggles humans will face learning to survive in a hostile, foreign environment are difficulties we need to face and solve before we venture further into the solar system. Learning how to live and work on the Moon involves both humans and machines, together, coping with an environment of low gravity, vacuum, thermal extremes, and hard radiation. We can design equipment to use and to protect us for short durations, but we need to understand how well these instruments and machines work on timescales of months and years. Using the Moon as a natural laboratory will teach us how to arrive, survive, and thrive on other worlds.

Besides survival, we also need to learn how to explore and study alien worlds. We have a vague idea that such an exploration template somehow involves both humans and robots, but how do they interact and work together and apart to yield the maximum benefit? As space destinations and objectives become more complex and dangerous, it makes good sense to use the Moon to learn how to properly conduct the serious business of exploration. Humans yearn to explore. By doing so, they acquire strategic knowledge that increases our odds for survival. Making new discoveries broadens the imagination and allows us to envision solutions to problems that might otherwise not have occurred to us. Practical experience on the Moon will serve us well as we begin humanity's movement into the universe.

10

Where Do We Go From Here?

Many of us working in or with NASA recognized that the 2004 Vision for Space Exploration (VSE) was a conceptual breakthrough. The goal of using off-planet resources to enable new capabilities in spaceflight was the fulcrum we needed to change our approach and direction to spaceflight. Making this change would open the door to a wide variety of previously unobtainable mission concepts and ideas.

In this book, I have shared my perspective on why and how the VSE was conceived, executed, and eventually terminated—a cautionary tale, if you will, but I hope an instructive one. Lessons drawn from this history can keep us from repeating similar mistakes and help us create a better American space program, one that moves humanity into the solar system by creating new opportunities and expanding, rather than consuming, wealth.

Because NASA's response to the VSE was to focus on the first human mission to Mars, they devised an Apollo-style architecture, reverting to the only successful operational template for planetary exploration with which the agency was familiar. This decision effectively derailed the incremental and sustainable approach for the extension of human reach into space, intended by the VSE. An Apollo-style mission to Mars remains a bridge too far fiscally, technically, and politically. The interpretation of a human Mars mission as the central goal for the agency ignored the considered work of the VSE architects and those of us who had worked with NASA in the immediate years following its announcement. Certainly I was not interested in participating in a new "Mission to Mars" paper study that was doomed to failure from the beginning. Many of us had already experienced this during the years of the Space Exploration Initiative (1989–92), an earlier attempt to re-create the Apollo zeitgeist.

The excitement that many of us felt at the beginning of the VSE came from the belief that those lessons about what did *not* work had been well learned and that a long overdue change in the template of spaceflight was upon us. We soon became disabused of such a notion. Although many in the space community understood both the possibilities and the pitfalls of the new effort, the dominant culture in both the agency and industry was wedded (and remains wedded) to the old template. As NASA reverted to their comfort zone, both the impression and the reality that this was about "repeating" our previous experience with the Moon—to regain the glory of Apollo—was cemented in many minds. This mindless calculus branded the lunar segment of the VSE with a "been there, done that" label, leading to the inevitable characterization that the Moon was both old hat and an unaffordable distraction.

Given that NASA was handed a new and challenging mission to go to the Moon on two previous occasions, the SEI in 1989 and the VSE in 2004—and both times they dropped the ball on implementing it—one might imagine that a new entity is needed to conduct human spaceflight for the federal government. This concept has not gone unvoiced by the community; no less an authority than Harrison "Jack" Schmitt, Apollo 17 astronaut and the first (and for now, the last) scientist to explore the Moon, has proposed that NASA be abolished and that a new agency be established to implement a long-range, strategic plan for human spaceflight.[1] Schmitt would reassign some NASA activities (aeronautics research, astronomy) to other agencies and retain within the new entity only the field centers critical to human spaceflight. The new space agency would take over existing infrastructure for these functions and maintain a minimal headquarters presence in Washington to preside over policy decisions.

I sympathize with Schmitt's frustration at the obtuseness and intransigence of the existing agency, but I think that his reformulation idea, while having much to commend it, is unlikely to be realized under normal circumstances. Too many entrenched interests, political as well as local, would be affected negatively by such a major reconstitution. However, some institutional crisis of confidence, a series of disasters or evidence of massive incompetence could produce the political momentum for radical change. This has happened in NASA's past—the Apollo 1 fire in 1967 resulted in a wholesale housecleaning of at least the upper management of the Apollo program, and the *Challenger* accident in 1986 likewise caused much soul-searching. The series of blunders in the early 1990s involving the faulty mirror of the Hubble Space Telescope and failure of two Mars missions led to calls for an agency shakeup. Each time, NASA was able to shrug off any significant institutional impacts, but in the midst of some future disaster, their bureaucratic luck may finally run out. However, the national mood seems primed lately to demand accountability in our government institutions and elected officials. Such a policy environment may yet result in a major reconfiguration of the civil space agency.

Flights to supply the International Space Station (ISS) using non-NASA "commercial" spacecraft are portrayed as a new goal and direction for space, even though the development of these new vehicles has been and will continue to be largely billed to the American taxpayer, as will be true of their operational costs. These days, transporting our astronauts to the ISS, the space station that we primarily designed, built, and paid for, requires that we pay the going rate to fly aboard a Russian Soyuz spacecraft or stay home. We find ourselves in the untenable situation of having a dysfunctional space program with no strategic direction. Our nation's dire financial situation is rapidly approaching crisis proportions. It is highly likely that future space budgets will be flat at best, but more probably, lower than current levels of funding.

Our current program direction, the promise of a human Mars mission—as yet unachievable, but perhaps doable 25 to 30 years down the road—remains the principal roadblock to implementing a workable program based on the use of off-planet resources. It doesn't have to be. It is highly likely that we will achieve our first human mission to Mars only through the use of pro-pellant produced on the Moon.[2] The Moon has much more to offer, both as a testing ground for advanced planetary surface systems and as a natural laboratory to learn the skills required for a new generation of planetary explorers. A realistic architecture for Mars incorporates and utilizes the valuable resources of the Moon.

Those who believe that we should proceed directly to Mars and bypass the Moon might consider the following. Martian gravity is twice as strong as the Moon's. With aerobraking, delta-v to the surface of Mars is roughly 1000 m/s, and ascent to orbit from the surface is about 5000 m/s. This means that you must bring an ascent or descent module with you to Mars; if you were to go to Mars with the intent to settle there, it would perforce be a one-way trip. On the Moon, we require roughly 2000 m/s up or down. This can be accommodated with a single-stage vehicle, meaning that we can reuse this spacecraft to enable continual travel between lunar orbit and the surface. Reusability enables an affordable solution to the problem of establishing an off-planet presence; travel back and forth to the surface coupled with an incremental buildup of the outpost on the Moon makes the creation of a permanent presence there possible in a manner that is not possible on Mars, where discarded, once-used pieces result in an expensive, unsustainable transportation architecture.

The current spaceflight template established 60 years ago is to custom-design and build spacecraft, then launch them on expendable vehicles: design, build, fly, use, and discard. Born of necessity, this operational model ensures that spacecraft are complex, expensive, and serve a limited lifetime. It demands that we launch everything we need from Earth—from the bottom of the deepest gravity well in the inner solar system—requiring significant energy (read "cost") to reach an intended destination. Until we change our national approach to the problem of spaceflight, we will remain mass- and power-limited, and therefore capability-limited in space. These necessary, expensive, and difficult goals are achievable under constrained budgets by taking small, affordable incremental steps that build on each other and work together to create a greater capability over time.

Nearly all of our modern space assets reside in the zone between Earth and Moon (cislunar space) and the difficulty of reaching low Earth orbit (LEO) limits our activities there. These cislunar satellites constitute the backbone of modern technical civilization and conduct critical societal functions such as communications, positioning, remote sensing, weather monitoring, and national strategic surveillance. The size and capability of such assets are limited by the size of the largest rocket that can launch a given payload and by their preordained operational lifetime. Our experience working with the space shuttle and ISS programs has demonstrated that people and machines working together, over time, can assemble and maintain space systems that can be made as large and operated for as long as desired. The problem is moving people and robots to these various points in cislunar space.

To become a spacefaring species, we must develop and possess freedom of movement and action, throughout cislunar space. Robotic missions show that the Moon's poles contain significant amounts of water ice, the most useable resource for humans in space. As a consumable, H_2O (water and oxygen) supports life. Used as shielding, water can protect people from cosmic radiation. Water is also a medium of energy storage; it can be dissociated into its component hydrogen and oxygen using electricity generated by sunlight and during local night or eclipse, these gases can be combined back into water to generate electricity. Finally, liquid hydrogen and oxygen are the most powerful chemical rocket propellant known, which opens the possibility for the Moon to become our first "offshore" coaling station in the sea of cislunar space.

Because the Moon is close, the time delay for a round-trip radio signal is less than three seconds. This gift of proximity makes it possible for machines, under the control of operators on Earth, to begin the initial work of establishing a demonstration resource processing facility on the Moon. Transit times to the Moon are as short as three days, and launch opportunities are always available. Some peaks and crater rims near the ice-rich lunar poles experience nearly constant sunlight, permitting the near-constant generation of electrical power with solar arrays. The individual pieces of equipment necessary to begin the harvesting of lunar ice are small and can be launched on small and medium-lift rockets. We can begin to install and operate a lunar polar resource extraction facility now, without waiting for the advent of new, heavy lift launch systems. A scaled, incremental approach to building a facility on the Moon can fit under nearly any budgetary envelope and offers numerous, intermediate milestones to document accomplishment and to map steady progress. Finally, the use of multiple, small steps to develop the Moon facilitates the participation of both international and commercial partners in creating a permanent space transportation system.

Making the Moon and cislunar space our next strategic goal in space solves many problems. It creates a near-term (decadal, not multidecadal) objective against which progress can be demonstrated and measured, inviting myriad ideas and participation. It can be built in incremental steps, tailored to be affordable under a wide variety of restrictive budget regimes. It creates a lasting infrastructure that allows people and machines access to all of the locations in cislunar space—the location of scientific, economic, and strategic assets. We will finally have laid the groundwork necessary to navigate past self-imposed roadblocks, thereby opening the solar system to exploration through the creation of a space transportation network that allows routine departure from, and return to, low Earth orbit.

Because we are dependent on space assets—the technology that controls, assists, and enhances so much of our daily lives—the current aimless direction of our civil space program not only endangers the agency's future but also jeopardizes critical national interests. Creating routine access to cislunar space will allow us to graduate from the "flags and footprints" model of human space travel to the creation, use, and control of a true, long-term spacefaring capability. We can do this in a manner that is scalable and thus affordable. It is the right direction for our civil space program in the new millennium.

Developing cislunar space and the Moon is a challenging but achievable goal. Although we are uncertain where this journey ultimately will take us, history records that humanity always gains knowledge and prospers when we expand our horizons. Using the Moon's resources to explore space and to live and prosper there will increase our chances for long-term survival and improve our quality of life. This great challenge holds the promise of breakthrough technologies and new discoveries that will ensure better futures for us all.

NOTES

1. Luna: Earth's Companion in Space

[1] See B. Brunner, *Moon: A Brief History* (New Haven: Yale University Press, 2010), for a readable compilation of the cultural influences of the Moon on humanity.

[2] D. J. Boorstin, *The Discoverers* (New York: Vintage, 1985).

[3] E. A. Whitaker, *Mapping and Naming the Moon: A History of Lunar Cartography and Nomenclature* (Cambridge: Cambridge University Press, 1999).

[4] See W. G. Hoyt, *Coon Mountain Controversies: Meteor Crater and the Development of Impact Theory* (Tucson: University of Arizona Press, 1987), for a lively recounting of this controversy.

[5] G. K. Gilbert, "The Origin of Hypotheses, Illustrated by the Discussion of a Topographic Problem," *Science* 3, no. 53 (1896): 1–15; http://www.sciencemag.org/content/3/53/1.extract.

[6] W. Ley, *Rockets, Missiles and Men in Space* (New York: Viking, 1966).

[7] D. E. Wilhelms, *To a Rocky Moon: A Geologist's History of Lunar Exploration* (Tucson: University of Arizona Press, 1993).

[8] Ibid.

[9] J. L. Powell, *Night Comes to the Cretaceous: Dinosaur Extinction and the Transformation of Modern Geology* (New York: W. H. Freeman, 1998).

[10] J. M. Logsdon, *After Apollo? Richard Nixon and the American Space Program* (New York: Palgrave Macmillan, 2015).

[11] H. L. Shipman, *Humans in Space: 21st Century Frontiers* (Plenum, New York, 1989). This prescient book presented a clear-eyed analysis of the conditions under which space settlement might be achieved (page 308, Table 9):

Possible Space Futures

		Will space industrialization work?	Will space industrialization work?
		yes	no
Can extraterrestrial resources be used to support humans in space?	yes	Full settlement 42%	Research and tourism 14%
Can extraterrestrial resources be used to support humans in space?	no	Robot mines, factories and labs 18%	Space science only 26%

[12] K. Ehricke, "Lunar Industrialization and Settlement: Birth of a Polyglobal Civilization," *Lunar Bases and Space Activities of the 21st Century* (Houston, TX: Lunar and Planetary Institute Press, 1985), 827–855; http://tinyurl.com/ob74goo.

[13] http://www.cislunarnext.org.

2. The Moon Conquered—and Abandoned

[1] R. B. Baldwin, *The Face of the Moon* (Chicago: University of Chicago Press, 1949).

[2] A. C. Clarke, *The Exploration of Space* (New York: Harper and Bros., 1951).

[3] H. C. Urey, *The Planets, Their Origin and Development* (New Haven: Yale University Press, 1952).

[4] W. G. Hoyt, *Coon Mountain Controversies: Meteor Crater and the Development of Impact Theory* (Tucson: University of Arizona Press, 1987)

[5] E. M. Shoemaker, Lunar Photogeologic Chart LPC 58 (1960), http://www.lpi.usra.edu/resources/mapcatalog/LunarPhotogeologicChart.

[6] W. von Braun et al., *Across the Space Frontier* (New York: Viking, 1952).

[7] W. Ley, *Rockets, Missiles and Men in Space* (New York: Viking, 1966).

[8] C. Murray and C. B. Cox, *Apollo: The Race to the Moon* (New York: Simon & Schuster, 1989).

[9] R. Zimmerman, *Genesis: The Story of Apollo 8* (New York: Four Walls Eight Windows, 1998).

[10] Space Task Group, "The Post-Apollo Space Program: Directions for the Future" (1969), http://www.hq.nasa.gov/office/pao/History/taskgrp.html.

[11] See A. Chaikin, *A Man on the Moon* (New York: Viking Press, 1994), for an excellent description of the explorations and adventures of the last three Apollo explorations.

[12] J. L. Powell, *Night Comes to the Cretaceous: Dinosaur Extinction and the Transformation of Modern Geology* (New York: W. H. Freeman, 1998).

[13] See N. L. Johnson, *The Soviet Reach for the Moon* (New York: Cosmos Books, 1995), and A. A. Siddiqi *Challenge to Apollo: The Soviet Union and the Space Race 1945–1974* (Washington, DC: NASA, 2000), for details on the Soviet lunar effort.

[14] Ibid.

[15] K. Adelman, *Reagan at Reykjavik: Forty-Eight Hours That Ended the Cold War* (New York: Broadside Books, 2014).

[16] http://www.spudislunarresources.com/Opinion_Editorial/Apollo_30_op-ed.htm.

[17] See D. Pettit, "The Tyranny of the Rocket Equation" (2011), http://www.nasa.gov/mission_pages/station/expeditions/expedition30/tryanny.html.

[18] B. G. Drake, ed., *Human Exploration of Mars Design Reference Mission 5.0*, NASA SP-2009–566 (2009), http://www.nasa.gov/pdf/373665main_NASA-SP-2009–566.pdf.

[19] http://en.wikipedia.org/wiki/Apollo_program#Program_cost.

3. After Apollo: A Return to the Moon?

[1] See M. D. Tribbe, *No Requiem for the Space Age: The Apollo Moon Landings and American Culture* (New York: Oxford University Press, 2014), for a discussion of the social criticism of the Apollo program.

[2] J. M. Logsdon, "The Space Shuttle: A Policy Failure," *Science* 232 (1986): 1099–1105; http://www.sciencemag.org/content/232/4754/1099.

[3] L. F. Belew, *Skylab: Our First Space Station*, NASA SP-400 (1977), http://history.nasa.gov/SP-400/contents.htm.

[4] D. R. Jenkins, *Space Shuttle: The History of the National Space Transportation System* (Stillwater, MN: Voyageur Press, 2002).

[5] E. C. Ezell and L. N. Ezell, *The Partnership: A History of the Apollo-Soyuz Test Project*, NASA SP-4209 (1978), http://www.hq.nasa.gov/office/pao/History/SP-4209/toc.htm.

[6] T. R. Heppenheimer, *The Space Shuttle Decision, 1972–1981* (Washington, DC: Smithsonian Institution Press, 2002).

[7] W. von Braun et al., *Across the Space Frontier* (New York: Viking, 1952).

[8] Ibid.

[9] H. E. McCurdy, *The Space Station Decision: Incremental Politics and Technological Choice* (Baltimore: Johns Hopkins University Press, 1990).

[10] http://www.astronautix.com/craft/otv.htm.

[11] W. W. Mendell, ed., *Lunar Bases and Space Activities of the 21st Century* (Houston, TX: Lunar and Planetary Institute Press, 1985).

[12] P. D. Spudis, "Lunar Resources: Unlocking the Space Frontier," *Ad Astra* 23, no. 2 (Summer 2011), http://www.nss.org/adastra/volume23/lunarresources.html.

[13] H. H. Schmitt, *Return to the Moon: Exploration, Enterprise, and Energy in the Human Settlement of Space* (New York: Praxis-Copernicus, 2006).

[14] See J. R. Arnold, "Ice in the Lunar Polar Regions," *Journal of Geophysical Research* 84 (1979): 5659–5668.

[15] http://history.nasa.gov/rogersrep/genindex.htm

[16] National Commission on Space (Paine Report), *Pioneering the Space Frontier* (New York: Bantam Books, 1986).

[17] S. K. Ride et al. (Ride Report), *Leadership and America's Future in Space* (Washington, DC: NASA, 1987).

[18] T. Hogan, *Mars Wars: The Rise and Fall of the Space Exploration Initiative*, NASA Special Publication SP-2007–4410 (2007), http://history.nasa.gov/sp4410.pdf.

[19] Ibid.

[20] NASA (90-Day Study), *Report of the 90-Day Study on Human Exploration of the Moon and Mars* (Washington, DC: NASA, 1989), http://history.nasa.gov/90_day_study.pdf

[21] Hogan, *Mars Wars.*

[22] See D. Day, "Aiming for Mars, Grounded on Earth," *The Space Review* (2004), http://www.thespacereview.com/article/106/2.

[23] Synthesis Group (Stafford Report), *America at the Threshold: The Space Exploration Initiative* (Washington DC: US Government Printing Office, 1991), http://www.lpi.usra.edu/lunar/strategies/Threshold.pdf.

[24] E. J. Chaisson, *The Hubble Wars* (New York: HarperCollins, 1994).

[25] Hogan, *Mars Wars*.

[26] D. R. Baucom, "The Rise and Fall of Brilliant Pebbles," *Journal of Social and Political Economic Studies* 29, no. 2 (2004): 143–190.

[27] H. E. McCurdy, *Faster, Better, Cheaper: Low-cost Innovation in the U.S. Space Program* (Baltimore: Johns Hopkins University Press, 2001).

[28] B. J. Butler, D. O. Muhleman, and M. A. Slade, "Mercury: Full Disk Radar Images and the Detection and Stability of Ice at the North Pole," *Journal of Geophysical Research* 98, E8 (1993): 15003–15023.

[29] S. Nozette, C. Lichtenberg, P. D. Spudis, R. Bonner, W. Ort, E. Malaret, M. Robinson, and E. M. Shoemaker, "The Clementine Bistatic Radar Experiment," *Science* 274 (1996): 1495–1498.

[30] D. B. J. Bussey, P. D. Spudis, and M. S. Robinson, "Illumination Conditions at the Lunar South Pole," *Geophysical Research Letters* 26, no. 9 (1999): 1187; D. B. J. Bussey, K. E. Fristad, P. M. Schenk, M. S. Robinson, and P. D. Spudis, "Constant Illumination at the Lunar North Pole," *Nature* 434 (2005): 842; http://en.wikipedia.org/wiki/Peak_of_eternal_light.

[31] McCurdy, *Faster, Better, Cheaper*, describes Clementine's effect on subsequent NASA programs.

4. Another Run at the Moon

[1] A short history of this program is available at the NASA Discovery Program web site: http://discovery.nasa.gov/lib/pdf/HistoricalDiscoveryProgramInformation.pdf.

[2] P. D. Spudis, "Ice on the Moon," *The Space Review* (2006), http://www.thespacereview.com/article/740/1.

[3] N. J. S. Stacy and D. B. Campbell, "A Search for Ice at the Lunar Poles," *Lunar and Planetary Science* XXVI (1995): 1672; http://www.lpi.usra.edu/meetings/lpsc1995/pdf/1672.pdf

[4] Gene Shoemaker was killed in an automobile accident in Australia in 1997.

[5] W. H. Lambright, *Why Mars: NASA and the Politics of Space Exploration* (Baltimore: Johns Hopkins University Press, 2014).

[6] K. Sawyer, *The Rock From Mars: A Detective Story on Two Planets* (New York: Random House, 2006).

[7] Ibid.

[8] http://www.astrobio.net/topic/solar-system/mars/deciphering-mars-follow-the-water.

[9] See B. Burrough, *Dragonfly: NASA and the Crisis Aboard Mir* (New York: HarperCollins, 1998).

[10] D. M. Harland and J. E. Catchpole, *Creating the International Space Station* (Berlin: Springer-Praxis, 2002).

[11] First Lunar Outpost (FLO), NASA-JSC (1992); http://www.nss.org/settlement/moon/FLO.html.

[12] J. K. Strickland (2011); http://www.nss.org/settlement/mars/AccessToMars.pdf.

[13] Fleshed out in a fully mature, corrected form in R. Zubrin and R. Wagner, *The Case for Mars: The Plan to Settle the Red Planet and Why We Must* (New York: Free Press, 1996).

[14] See Chapter 3, note 29.

[15] A. Matsuoka and C. Russell, eds., *The Kaguya Mission to the Moon* (Berlin: Springer, 2011).

[16] See B. R. Blair, "Quantitative Approaches to Lunar Economic Analysis" (2009), for some of the conclusions of this work: http://tinyurl.com/pr6ktmd.

[17] This sad story is well told in M. Cabbage and W. Harwood, *Comm Check: The Final Flight of Shuttle Columbia* (New York: Free Press, 2008).

[18] Details of this long policymaking process are described in F. Sietzen and K. L. Cowing, *New Moon Rising: The Making of America's New Space Vision and the Remaking of NASA* (Burlington, ON: Apogee Books, 2004).

[19] Ibid.

[20] See these documents at the Klaus Heiss files, http://www.spudislunarresources.com/klaus.htm.

[21] Gold Team study products are unpublished; summaries presented in this book are taken from files in my collection.

[22] Columbia Accident Investigation Board (CAIB), *Report of the Columbia Accident Investigation Board* (Washington, DC: NASA, 2003); http://www.nasa.gov/columbia/home/CAIB_Vol1.html.

[23] "Bush May Announce Return to Moon at Kitty Hawk," *Space Daily*, October 29, 2003; http://www.spacedaily.com/news/beyondleo-03a.html.

[24] Sietzen and Cowing, *New Moon Rising*.

[25] G. W. Bush, "A Renewed Spirit of Discovery" (2004), White House; http://www.spaceref.com/news/viewpr.html?pid=13404.

5. Implementing the Vision

[1] G. W. Bush, "A Renewed Spirit of Discovery" (2004), White House; www.spaceref.com/news/viewpr.html?pid=13404.

[2] I trace this evolution, including excerpts from the internal Headquarters Red team/Blue Team activity in a presentation, "The Vision and the Mission," at http://tinyurl.com/aqer2t.

[3] http://history.nasa.gov/DPT/DPT.htm.

[4] Some involved with the DPT contend that it was a capability-driven (rather than a destination driven) strategic planning effort. See H. Thronson, "NASA's Decadal Planning Team (DPT) and the NASA Exploration Team (NEXT)" (2014), http://history.nasa.gov/DPT/thronson.pdf. The Mars and Quest for Life fixations of the agency were never far beneath the surface. See especially this presentation at the National Academy of Engineering web site: http://www.naefrontiers.org/File.aspx?id=22013. It reflects agency thinking on this topic.

[5] http://tinyurl.com/nagq2tn

[6] http://tinyurl.com/aqer2t.

[7] Comments at last public meeting, Aldridge Commission, June 2004: http://tinyurl.com/krlzdpz.

[8] President's Commission on the Implementation of Space Exploration Policy (Aldridge Report), *Journey to Inspire, Innovate and Discover* (Washington, DC: US Government Printing Office, 2004).

[9] Objectives/Requirements Definition Team (ORDT) for 2008 Lunar Reconnaissance Orbiter, http://tinyurl.com/k6ffpdx.

[10] Mini-RF imaging radar instruments: http://tinyurl.com/n8fnlvl.

[11] I wrote a few articles recounting my experiences during the Chandrayaan-1 mission for *Air & Space* magazine: http://tinyurl.com/koqzdkl, http://tinyurl.com/lhul4b6, http://tinyurl.com/lcdge3v.

[12] http://tinyurl.com/oh8ktrh.

[13] http://tinyurl.com/nk78bk8.

[14] http://www.space.com/15406-blue-origin-private-spacecraft-infographic.html.

[15] http://www.nasa.gov/about/highlights/griffin_bio.html.

[16] http://tinyurl.com/pnvgjyr.

[17] http://www.nasa.gov/exploration/news/ESAS_report.html.

[18] Columbia Accident Investigation Board (CAIB), *Report of the Columbia Accident Investigation Board* (Washington, DC: NASA, 2003); http://www.nasa.gov/columbia/home/CAIB_Vol1.html.

[19] J. Marburger, Keynote Address, 44th Goddard Memorial Symposium (2006), http://www.spaceref.com/news/viewsr.html?pid=19999.

[20] http://www.spaceref.com/news/viewpr.html?pid=13404.

[21] C. Bergin, "Digging Deeper into NASA's Moon Plans," *NASA Spaceflight*, December 4, 2006, http://www.nasaspaceflight.com/2006/12/digging-deeper-into-nasas-moon-plans. See also http://www.nasa.gov/pdf/163896main_LAT_GES_1204.pdf.

[22] D. Beattie, "Just How Full of Opportunity Is the Moon?" *The Space Review* (2007), http://www.thespacereview.com/article/804/1.

[23] http://lcross.arc.nasa.gov/index.htm.

[24] D. Shiga, "NASA May Abandon Plans for Moon Base," *New Scientist* (2009), http://tinyurl.com/oz54p2x.

[25] http://tinyurl.com/koqzdkl.

[26] P. D. Spudis et al., "Initial Results for the North Pole of the Moon from Mini-SAR, Chandrayaan-1," *Geophysical Research Letters* 37 (2010), http://tinyurl.com/pl5hjh6.

[27] P. D. Spudis, "Return to the Moon: Outpost or Sorties?" *Air & Space* (2009), http://tinyurl.com/mkhtflz.

[28] http://tinyurl.com/lyg4du.

[29] N. R. Augustine et al. (Augustine Report), *Advisory Committee on the Future of the U.S. Space Program* (Washington, DC: NASA, 1990), http://history.nasa.gov/augustine/racfup1.htm.

[30] Review of Human Spaceflight Plans Committee (Augustine Committee), *Seeking a Human Spaceflight Program Worthy of a Great Nation* (Washington, DC: NASA, 2010), http://www.nss.org/resources/library/spacepolicy/HSF_Cmte_FinalReport.pdf.

[31] http://history.nasa.gov/DPT/DPT.htm.

[32] In May 2013, Bolden was quoted as saying, "We need to try and get all of us on to the same sheet of music in terms of the roadmap. [If we] have someone in the next administration who could take us back to a human lunar mission, it's all over, we will go back to square one." http://www.nasaspaceflight.com/2013/05/return-moon-send-nasa-square-one-bolden.

[33] Obama space policy speech, NASA Kennedy Space Center, April 15, 2010; http://www.nasa.gov/news/media/trans/obama_ksc_trans.html.

[34] http://www.airspacemag.com/daily-planet/the-authorized-version-156372809.

[35] http://www.nasa.gov/exploration/systems/sls.

[36] http://www.spudislunarresources.com/blog/lets-haul-asteroids.

[37] http://en.wikipedia.org/wiki/Budget_of_NASA. Projected budget amounts for NASA without the VSE and projected augmented budget with VSE (Figure 5.2) are from a presentation by Administrator Sean O'Keefe, available at http://tinyurl.com/ns5t3j3.

6. Why? Three Reasons the Moon Is Important

[1] A summary of the six themes developed at this workshop can be seen on a poster: http://www.nss.org/settlement/moon/NASAwhymoon.pdf.

[2] See J. Lovell and J. Kluger, *Lost Moon: The Perilous Journey of Apollo 13* (Boston: Houghton Mifflin, 1994).

[3] See National Research Council, *The Scientific Context for Exploration of the Moon* (Washington, DC: National Academies Press, 2007), http://www.nap.edu/openbook.php?record_id=11954.

[4] See J. L. Powell, *Night Comes to the Cretaceous: Dinosaur Extinction and the Transformation of Modern Geology* (New York: W. H. Freeman, 1998).

[5] P. D. Spudis, "Lunar Resources: Unlocking the Space Frontier," *Ad Astra* 23, wno. 2 (Summer 2011), http://www.nss.org/adastra/volume23/lunarresources.html.

[6] L. A. Taylor and T. T. Meek, "Microwave Sintering of Lunar Soil: Properties, Theory and Practice," *Journal of Aerospace Engineering* 18 (2005): 188–196; http://www.isruinfo.com/docs/microwave_sintering_of_lunar_soil.pdf.

[7] D. B. J. Bussey, P. D. Spudis, and M. S. Robinson, "Illumination Conditions at the Lunar South Pole," *Geophysical Research Letters* 26, no. 9 (1999): 1187; D. B. J. Bussey, K. E. Fristad, P. M. Schenk, M. S. Robinson, and P. D. Spudis, "Constant Illumination at the Lunar North Pole," *Nature* 434 (2005): 842.

[8] Discussed in detail in H. H. Schmitt, *Return to the Moon: Exploration, Enterprise, and Energy in the Human Settlement of Space* (New York: Praxis-Copernicus, 2006).

[9] For example, see http://www.planetary.org/blogs/bill-nye/20130710-the-goal-is-mars.html.

[10] http://www.nss.org/settlement/mars/AccessToMars.pdf.

[11] See the current NASA Mars Design Reference Mission 5.0, http://www.nasa.gov/pdf/373665main_NASA-SP-2009-566.pdf.

[12] Space Task Group, "The Post-Apollo Space Program: Directions for the Future" (1969), White House, http://www.hq.nasa.gov/office/pao/History/taskgrp.html; http://www.spaceref.com/news/viewpr.html?pid=13404.

[13] See, for example, W. W. Mendell, "Meditations on the New Space Vision: The Moon as a Stepping Stone to Mars," *Acta Astronautica* 57 (2005): 676–683; http://www.ncbi.nlm.nih.gov/pubmed/16010766.

[14] R. Launius, "Exploding the Myth of Popular Support for Project Apollo" (2010), http://tinyurl.com/k9k9jt4.

7. How? Things We Should Have Been Doing

[1] A good introduction to general astronautics for the nontechnical reader can be found in G. Swinerd, *How Spacecraft Fly* (New York: Copernicus-Springer, 2008).

[2] For a good, nontechnical explanation, see D. Pettit, "The Tyranny of the Rocket Equation" (2001), http://www.nasa.gov/mission_pages/station/expeditions/expedition30/tyranny.html.

[3] Ibid.

[4] http://www.fai.org/icare-records/100km-altitude-boundary-for-astronautics.

[5] See B. F. Kutter et al., "A Practical Affordable Propellant Depot in Space Based on ULA's Flight Experience," *Space 2008*, AIAA 2008–7644 (2008), http://tinyurl.com/l2ndspu.

[6] http://tinyurl.com/qhw5nsu.

[7] J. M. Logsdon, "The Space Shuttle: A Policy Failure," *Science* 232 (1986): 1099–1105; http://www.sciencemag.org/content/232/4754/1099.

[8] http://www.nasa.gov/exploration/systems/orion/index.html; http://www.nasa.gov/exploration/systems/sls/index.html.

[9] http://www.ulalaunch.com/products_deltaiv.aspx.

[10] http://www.spacex.com/falcon-heavy.

[11] P. D. Spudis and A. R. Lavoie, "Using the Resources of the Moon to Create a Permanent Cislunar Space Faring System," *Space 2011*, AIAA, 2011–7185 (2011); http://www.spudislunarresources.com/Bibliography/p/102.pdf.

[12] http://solarsystem.nasa.gov/rps/rtg.cfm.

[13] Spudis and Lavoie, "Using the Resources of the Moon."

[14] http://en.wikipedia.org/wiki/Budget_of_NASA.

[15] See D. A. Day, "Whispers in the Echo Chamber," *The Space Review* (2004), http://www.thespacereview.com/article/119/1.

[16] Spudis and Lavoie, "Using the Resources of the Moon."

[17] Review of Human Spaceflight Plans Committee (Augustine Committee), *Seeking a Human Spaceflight Program Worthy of a Great Nation* (Washington, DC: US Government Printing Office, 2010), http://www.nss.org/resources/library/spacepolicy/HSF_Cmte_FinalReport.pdf.

8. If Not Now, When? If Not Us, Who?

[1] See J. M. Logsdon, *John F. Kennedy and the Race to the Moon* (New York: Palgrave Macmillan, 2010); on desalination, see http://www.desalination.com/museum/office-saline-water-desal-rd-funding-usa.

[2] See R. Rhodes, *The Making of the Atomic Bomb* (New York: Simon & Schuster, 1986).

[3] J. Marburger, Keynote Address, 44th Goddard Memorial Symposium (2006), http://www.spaceref.com/news/viewsr.html?pid=19999.

[4] http://history.nasa.gov/spaceact.html.

[5] P. D. Spudis, "Faded Flags on the Moon," *Air & Space* (July 19, 2011), http://tinyurl.com/oro7ome. Bolden's quote comes from the *Denver Post* of February 12, 2010.

[6] *Science in Orbit. The Shuttle and Spacelab Experience 1981–1986*, NASA NP-119, Chapter 7; http://history.nasa.gov/NP-119/ch7.htm.

[7] http://orbitaldebris.jsc.nasa.gov/index.html.

[8] R. Ridenoure, "Beyond GEO, Commercially: 15 Years and Counting," *The Space Review* (2013), http://www.thespacereview.com/article/2295/1.

[9] http://www.spudislunarrcsources.com/Opinion_Editorial/Apollo_30_op-ed.htm.

[10] http://www.dod.mil/pubs/space20010111.pdf.

[11] http://freebeacon.com/national-security/china-launches-three-asat-satellites.

[12] http://www.spudislunarresources.com/blog/china-in-space.

[13] President J. F. Kennedy to Joint Session of Congress, May 25, 1961: "We go into space because whatever mankind must undertake, free men must fully share." http://www.nasa.gov/vision/space/features/jfk_speech_text.html.

[14] See the literature cited at http://www.hq.nasa.gov/office/hqlibrary/pathfinders/opinion.htm.

[15] http://lunar.xprize.org.

[16] http://ansari.xprize.org.

[17] http://www.unoosa.org/oosa/SpaceLaw/outerspt.html.

9. A Visit to the Future Moon

[1] http://tinyurl.com/knvua7l.

[2] http://www.nasa.gov/topics/moonmars/features/moon20090924.html.

[3] mars.nasa.gov/msl/mission/rover.

[4] http://www.nasa.gov/mission_pages/ladee/main.

[5] P. D. Spudis and A. R. Lavoie, "Using the Resources of the Moon to Create a Permanent Cislunar Space Faring System," *Space 2011*, AIAA, 2011–7185 (2011), http://www.spudislunarresources.com/Bibliography/p/102.pdf.

[6] http://www.space.com/18694-moon-dirt-3d-printing-lunar-base.html.

[7] http://www.nss.org/settlement/ColoniesInSpace/colonies_chap06.html.

[8] http://tinyurl.com/8zf4m5.

[9] http://www3.nd.edu/~cneal/Lunar-L/Moon-as-a-tape-recorder.pdf

[10] P. D. Spudis and G. J. Taylor, "The Roles of Humans and Robots as Field Geologists on the Moon," in *2nd Conference on Lunar Bases and Space Activities of 21st Century*, ed. W. Mendell, NASA Conference Publications 3166 (1992), 1:307–313. http://tinyurl.com/q44zh36.

[11] http://www.nss.org/settlement/ssp.

[12] http://www.spaceagepub.com/pdfs/Ignatiev.pdf.

[13] D. R. Criswell, "Solar Power via the Moon," *The Industrial Physicist*, April/May 2002, http://tinyurl.com/pmnelod.

[14] H. H. Schmitt, *Return to the Moon: Exploration, Enterprise, and Energy in the Human Settlement of Space* (New York: Praxis-Copernicus, 2006).

[15] http://lcross.arc.nasa.gov/observation.htm.

[16] I discuss the "Flexible Path" idea here: http://tinyurl.com/ok8wsvv.

10. Where Do We Go From Here?

[1] H. H. Schmitt, "Space Policy and the Constitution" (2011), Americasuncommonsense.com, http://tinyurl.com/luohafp.

[2] "ISRU and the Critical Path to Mars" (2013), Spudis Lunar Resources Blog, http://tinyurl.com/pw53hm2.

A LUNAR
LIBRARY

The literature of the Moon is enormous, and the following list makes no claim to completeness. These are books or sources that I have found to be important in one way or another. Several books deal with space topics other than the Moon; they are included because they are relevant to understanding the issues raised here. All books have their own bibliographies that will allow you to explore their topics in greater depth.

Two Essential Books
Both of these books can be downloaded at no cost on the Internet:

Heiken, G. H., D. T. Vaniman, and B. M. French, eds. 1991. *The Lunar Sourcebook: A User's Guide to the Moon*. Cambridge: Cambridge University Press. http://www.lpi.usra.edu/publications/books/lunar_sourcebook.
   The definitive reference book on the Moon, written by more than thirty active and former lunar scientists. Particularly thorough on lunar rocks and soils and nicely complements Wilhelms's book (below). Written for the lay reader, but does not flinch on technical concepts.

Wilhelms, D. E. 1987. *The Geologic History of the Moon*. US Geological Survey Professional Paper 1348. Washington, DC: US Government Printing Office. http://ser.sese.asu.edu/GHM.
   The historical geology of the Moon, written by one of the premier lunar geologists and historians of lunar science. The book is well written and illustrated. A cogent summary of our understanding of the Moon from the stratigraphic perspective.

History Lesson
Chaikin, A. 1994. *A Man on the Moon*. New York: Viking Press.
   Deals with Apollo from the astronauts' perspective. Very well done and interesting; covers both science and operations. Basis for the HBO TV series *From the Earth to the Moon*.

Collins, M. 1974. *Carrying The Fire: An Astronaut's Journeys*. New York: Farrar, Straus & Giroux.
>   The best book by any astronaut, even if he does dislike geology! Fascinating, funny, and profound. Read this book to get a real feel for what going to the Moon was like.

Compton, W. D. 1989. *Where No Man Has Gone Before: A History of Apollo Lunar Exploration Missions*. NASA Special Publication 4214. Washington, DC: US Government Printing Office. http://www.lpi.usra.edu/lunar/documents/NTRS/collection3/NASA_SP_4214.pdf.
>   The "official" NASA history of Apollo lunar exploration. Takes the stance that lunar flight was primarily a difficult engineering task, made even more difficult by continuously complaining scientists (a view with which I can sympathize). Complements the scientific viewpoint by Wilhelms's *To a Rocky Moon*.

Harland, D. M. 2008. *Exploring the Moon: The Apollo Expeditions*. Berlin: Springer-Praxis.
>   Superb, well illustrated, and comprehensive history of the Apollo missions, their goals, and the events of the program.

Heiken, G., and E. Jones. 2007. *On the Moon: The Apollo Journals*. Berlin: Springer-Praxis. http://www.hq.nasa.gov/office/pao/History/alsj/frame.html.
>   Book version of the famous Apollo Lunar Surface Journal, the complete online transcripts of the Apollo explorations of the Moon. Book summarizes the principal exploration and discoveries of each mission.

Hoyt, W. G. 1987. *Coon Mountain Controversies: Meteor Crater and the Development of Impact Theory*. Tucson: University of Arizona Press.
>   An exhaustive history of the study of Meteor Crater, Arizona, including much on the debate about the craters of the Moon. Highly recommended.

Johnson, N. L. 1995. *The Soviet Reach for the Moon*. New York: Cosmos Books. http://www.lpi.usra.edu/publications/books/sovietReach/index.pdf.
>   Brief history of the Soviet lunar program, with emphasis on the building and fate of the N-1 superbooster.

Murray, C., and C. B. Cox. 1989. *Apollo: The Race to the Moon*. New York: Simon & Schuster.
>   My favorite book about Apollo. Wonderfully told engineering side of the story, including a nail-biting account of the near-disaster we almost had during the first landing on the Moon. Captures the excitement of the early days like no other book.

Sawyer, K. 2006. *The Rock from Mars: A Detective Story on Two Planets*. New York: Random House.

> The saga of ALH84001, the famous meteorite from Mars in which evidence supposedly was found for ancient fossils. Interesting on the political fallout from the discovery, which was considerable.

Siddiqi, A. A. 2000. *Challenge to Apollo: The Soviet Union and the Space Race 1945–1974*. NASA SP-2000–4408. Washington, DC: NASA. http://history.nasa.gov/SP-4408pt1.pdf and http://history.nasa.gov/SP-4408pt2.pdf.

> Massive and comprehensive history of the Soviet space program. The definitive work.

Wilhelms, D. E. 1993. *To a Rocky Moon: A Geologist's History of Lunar Exploration*. Tucson: University of Arizona Press. http://www.lpi.usra.edu/publications/books/rockyMoon.

> The clearest, most complete account of the history of lunar science in the space age. Weak on the early phases (which are well covered in the books by Hoyt and by Sheehan and Dobbins), but unsurpassed for lunar science starting with Baldwin and including geological mapping, astronaut training, and site selection for the Apollo missions.

Wolfe, T. 1979. *The Right Stuff*. New York: Farrar, Straus & Giroux.

> The greatest book about the space program, even though space is actually a marginal part of Wolfe's story. The quintessence of America in spirit and substance, all the more startling in its contrast to the present space program and NASA.

Space Policy and Program History

Catchpole, J. E. 2008. *The International Space Station: Building for the Future*. Berlin: Springer-Praxis. Harland, D. M., and J. E. Catchpole. 2002. *Creating the International Space Station*. Berlin: Springer-Praxis.

> Comprehensive history of the space station program and operations through its construction and initial operations.

Heppenheimer, T. R. 2002. *The Space Shuttle Decision 1965–1972*. Washington, DC: Smithsonian Institution Press.

———. *The Space Shuttle Decision, 1972–1981*. Washington, DC: Smithsonian Institution Press.

> The "official" NASA history of the design and construction of the space shuttle, ending with its first flight in 1981. Some who worked in the program have told me that the early history is somewhat distorted.

Hogan, T. 2007. *Mars Wars: The Rise and Fall of the Space Exploration Initiative*. NASA Special Publication SP-2007–4410. Washington, DC: NASA. http://history.nasa.gov/sp4410.pdf
 Brief, superficial history of the declaration and fate of President George H. W. Bush's Human (Space) Exploration Initiative. Completely misses the Moon-Mars controversy, which arguably helped cause the demise of SEI.

Jenkins, D. R. 2002. *Space Shuttle: The History of the National Space Transportation System*. Stillwater, MN: Voyageur Press.
 Comprehensive and well-written book on the origins, building, and flights of the space shuttle. Nicely illustrated.

Kitmacher, G. H. 2010. *Reference Guide to the International Space Station: Assembly Complete Edition*. NASA NP-2010–09–682-HQ. Washington, DC: NASA. http://www.nasa.gov/pdf/508318main_ISS_ref_guide_nov2010.pdf.
 Beautifully illustrated book showing how the ISS was assembled and is operated. The definitive work; I did not fully understand the station until I read this book.

Logsdon, J. M. 2010. *John F. Kennedy and the Race to the Moon*. New York: Palgrave Macmillan.

———. 2015. *After Apollo? Richard Nixon and the American Space Program*. New York: Palgrave Macmillan.
 Two pieces by the dean of space policy history. Logsdon wrote the definitive work on the JFK Moon decision and aims to do the same for Nixon and the space shuttle, with somewhat less success.

McCurdy, H. E. 1990. *The Space Station Decision: Incremental Politics and Technological Choice*. Baltimore: Johns Hopkins University Press.
 Good history of the policy choices made during the design of *Freedom*, although ending before its existential crisis and subsequent rebirth as the International Space Station in the 1990s.

McDougall, W. A. 1985. *The Heavens and the Earth: A Political History of the Space Age*. New York: Basic Books.
 Exhaustive study of the politics of the space program and government technology research in general. Emphasis on the early Sputnik days.

Schmitt, H. H. 2006. *Return to the Moon: Exploration, Enterprise, and Energy in the Human Settlement of Space*. New York: Praxis-Copernicus.
 Mostly deals with the programmatic aspects of lunar return, focusing on the mining of helium-3. Jack Schmitt is the only professional scientist to have walked on the Moon.

Shipman, H. L. 1989. *Humans in Space: 21st Century Frontiers*. New York: Plenum.
Insightful, prophetic book that correctly identified the need to develop
and use the resources of space to create new capabilities.

Sietzen, F., and K. L. Cowing. 2004. *New Moon Rising: The Making of America's
New Space Vision and the Remaking of NASA*. Burlington, ON: Apogee Books.
The sole work on the origins of the Vision for Space Exploration policy. The
authors had access to several inside sources, making this an invaluable resource,
although it possesses the drawbacks of being an "instant history" effort.

Tribbe, M. D. 2014. *No Requiem for the Space Age. The Apollo Moon
Landings and American Culture*. New York: Oxford University Press.
Annoying book about the "whiners of Apollo," those "experts" who continually
complained about and denigrated the effort to go to the Moon throughout
the 1960s. A must-read, but take Dramamine before plunging in.

Zubrin, R., and R. Wagner. 1996. *The Case for Mars: The Plan to Settle
the Red Planet and Why We Must*. New York: Free Press.
The definitive exposition of the "Mars Direct" architecture by its originator.

Major Committee Reports on Space Policy

These are presented in chronological order, without comment. All may be accessed and read free on the Internet. They contain the good, the bad, and the ugly of space policy.

Space Task Group. 1969. *The Post-Apollo Space Program: Directions for the Future*.
Washington, DC: NASA. http://www.hq.nasa.gov/office/pao/History/taskgrp.html.

National Commission on Space (Paine Report). 1986. *Pioneering the Space Frontier*.
New York: Bantam Books. http://history.nasa.gov/painerep/begin.html.

Ride, S. K., et al. (Ride Report). 1987. *Leadership and America's Future in Space*.
Washington, DC: NASA. http://history.nasa.gov/riderep/main.PDF.

NASA (90-Day Study). 1989. *Report of the 90-Day Study on Human Exploration of the
Moon and Mars*. Washington, DC: NASA. http://history.nasa.gov/90_day_study.pdf

Augustine, N. R., et al. (Augustine Report). 1990. *Advisory
Committee on the Future of the U.S. Space Program*. Washington,
DC: NASA. http://history.nasa.gov/augustine/racfup1.htm.

Synthesis Group (Stafford Report). 1991. *America at the Threshold: The Space Exploration
Initiative*. Washington, DC: NASA. http://www.lpi.usra.edu/lunar/strategies/Threshold.pdf

Commission to Assess United States National Security Space Management and Organization (Rumsfeld Commission). 2001. *Report of the Commission to Assess United States National Security Space Management and Organization*. Washington, DC: US Department of Defense. http://www.dod.mil/pubs/space20010111.pdf.

Columbia Accident Investigation Board (CAIB). 2003. *Report of the Columbia Accident Investigation Board*. Washington, DC: NASA. http://www.nasa.gov/columbia/home/CAIB_Vol1.html.

President's Commission on the Implementation of Space Exploration Policy (Aldridge Report). 2004. *Journey to Inspire, Innovate and Discover*. Washington, DC: US Government Printing Office. http://www.nss.org/resources/library/spacepolicy/2004-AldridgeCommissionReport.pdf.

National Research Council. 2007. *The Scientific Context for Exploration of the Moon*. Washington, DC: National Academies Press. http://www.nap.edu/openbook.php?record_id=11954.

Review of Human Spaceflight Plans Committee (Augustine Committee). 2010. *Seeking a Human Spaceflight Program Worthy of a Great Nation*. Washington, DC: US Government Printing Office. http://www.nss.org/resources/library/spacepolicy/HSF_Cmte_FinalReport.pdf.

National Research Council. 2014. *Pathways to Exploration: Rationales and Approaches for a U.S. Program of Human Exploration*. Washington, DC: National Academies Press. http://www.nap.edu/openbook.php?record_id=18801.

Lunar Classics

Baldwin, R. B. 1949. *The Face of the Moon*. Chicago: University of Chicago Press.
    The study by Baldwin that got it all so right, so early. This book inspired Harold Urey's interest in the Moon and greatly influenced many early lunar scientists.

Hartmann, W. K., R. J. Phillips, and G. J. Taylor, eds. 1986. *Origin of the Moon*. Houston, TX: Lunar and Planetary Institute Press.
    http://www.lpi.usra.edu/publications/books/origin-of-the-moon.
    Proceedings of the great Kona Moon origin conference and hence, the definitive statement of the giant impact model for the origin of the Moon. Review papers by Wood, Drake, and Hood are particularly worthy; also see the history of the study of lunar origin by Brush.

Mendell, W. W., ed. 1985. *Lunar Bases and Space Activities of the 21st Century*. Houston, TX: Lunar and Planetary Institute Press. http://www.lpi.usra.edu/publications/books/lunar_bases.

————. 1992. *Second Conference on Lunar Bases and Space Activities of the 21st Century*. Washington, DC: NASA. http://www.nss.org/settlement/moon/library/lunar2.htm.
    The proceedings of two conferences in 1984 and 1988. Great fun. A
    collection of wild fantasies about the advent of another Apollo program,
    come to save us all from the purgatory of space mediocrity.

Mutch, T. A. 1970. *The Geology of the Moon: A Stratigraphic View*. Princeton: Princeton University Press.
    Wonderfully written and illustrated account of the stratigraphy (layered
    rocks) of the Moon. Although it is five decades old, many of the basic
    concepts (e.g., mapping relative ages) it describes remain current.

Schultz, P. H. 1976. *Moon Morphology: Interpretations based on Lunar Orbiter Photography*. Austin: University of Texas Press.
    Massive compilation of Lunar Orbiter images of just about every imaginable lunar
    feature, classified by type of landform. Images are well reproduced on quality paper.

Readable, Reliable Popular Accounts of Lunar Science and Exploration

Cortwright, E. M., ed. 1975. *Apollo Expeditions to the Moon*. NASA Special Publication 350. Washington, DC: US Government Printing Office. http://history.nasa.gov/SP-350/toc.html.
    A collection of essays on all aspects of the Apollo program, from booster rockets to
    lunar science, written by participants. Illustrated with many color photographs.

Crotts, A. 2014. *The New Moon: Water, Exploration and Future Habitation*. New York: Cambridge University Press.
    Massive review of recent lunar exploration results, with considerable
    (perhaps too much) attention paid to Lunar Transient Phenomena.

Lewis, J., M. S. Matthews, and M. L. Guerrieri, eds. 1993. *Resources of Near-Earth Space*. Tucson: University of Arizona Press. http://www.uapress.arizona.edu/onlinebks/ResourcesNearEarthSpace/contents.php.
    Compilation of review papers covering the material resources of space, focusing on
    the Moon and near-Earth objects. Written before the discovery of lunar polar ice.

Light, M. 1999. *Full Moon*. New York: Knopf. http://www.michaellight.net/fm-intro.
Magnificent coffee-table book of Apollo photographs.

Masursky, H., G. W. Colton, and F. El-Baz, eds. 1978. *Apollo over the Moon:
A View From Orbit*. NASA Special Publication 362. Washington, DC: US
Government Printing Office. http://history.nasa.gov/SP-362/contents.htm.
Collection of the best photographs taken from lunar orbit during the Apollo missions,
each one presented with a geologically oriented caption by a relevant expert.

Powell, J. L. 1998. *Night Comes to the Cretaceous: Dinosaur Extinction and
the Transformation of Modern Geology*. New York: W. H. Freeman.
Accessible account of the development and path of the revolution in geology
caused by the recognition that a giant impact 65 million years ago caused
the extinction of many species, including most famously, the dinosaurs.

Wingo, D. 2004. *Moonrush: Improving Life on Earth with the
Moon's Resources*. Burlington, ON: Apogee Books.
Discusses the concept of finding large amounts of platinum-group metals on the
Moon to serve a hydrogen-based energy economy on Earth. I have some technical
issues with this idea but agree that the Moon can serve the terrestrial economy.

Wood, C. A. 2003. *The Modern Moon: A Personal
View*. Cambridge, MA: Sky Publishing.
Nicely illustrated tour of the near side of the Moon for the
amateur astronomer, punctuated by some brief geological
narratives and anecdotal stories of various lunar scientists.

Lunar Atlases and Maps

Bowker, D. E., and J. K. Hughes. 1971. *Lunar Orbiter Photographic Atlas of
the Moon*. NASA Special Publication 206. Washington, DC: US Government
Printing Office. http://www.lpi.usra.edu/resources/lunar_orbiter.
The definitive collection of Lunar Orbiter pictures, showing almost the entire lunar
surface, both near and far sides. Its value is somewhat hampered by relatively poor
reproduction of some of the photographs. Now available in an online edition.

Bussey, B., and P. D. Spudis. 2012. *The Clementine Atlas of the Moon*.
Revised edition. Cambridge: Cambridge University Press.
The best and most comprehensive atlas of the Moon (if I do say so
myself), showing the surface and nomenclature of the entire lunar
surface at a consistent scale and degree of detail. Includes a brief
history and description of the findings of the Clementine mission,
which revolutionized our understanding of the Moon.

Hare, T. M., R. K. Hayward, J. S. Blue, and B. A. Archinal. 2015. *Image Mosaic and Topographic Map of the Moon*. US Geological Survey Scientific Investigations Map 3316. Washington, DC: US Geological Survey. http://dx.doi.org/10.3133/sim3316.
   The current "official" version of the USGS maps based on images and topographic data returned by the Lunar Reconnaissance Orbiter mission. Topographic map (sheet 2) is virtually worthless because of poor selections for color rendering.

National Geographic Society. 1976. *The Earth's Moon*. Second edition. Washington, DC: National Geographic Society.
   The best map of the Moon, showing both near and far sides (with major feature names) on a single sheet at a scale of 1.10,000,000. Margins are filled with fascinating facts and drawings about the Moon and an index of named formations.

Rükl, A. 1990. *Hamlyn Atlas of the Moon*. London: Hamlyn.
   Excellent atlas of the near side of the Moon, particularly useful for amateur astronomers and observers. Each map in the atlas gives a brief entry on the people for whom craters were named.

Stooke, P. J. 2007. *The International Atlas of Lunar Exploration*. Cambridge: Cambridge University Press.
   Compilation of maps showing the results of all lunar missions to date, at a variety of scales. Essential for the true lunar fanatic.

Whitaker, E. A. 1999. *Mapping and Naming the Moon: A History of Lunar Cartography and Nomenclature*. Cambridge: Cambridge University Press.
   The history of the mapping of the Moon by one of the great scholars of that field. Definitive and authoritative.

Moon Lore and Cultural History

Brunner, B. 2010. *Moon: A Brief History*. New Haven: Yale University Press.
   A collection of miscellany, myths, lore, and legends dealing with the Moon. Entertaining and fast paced.

Montgomery, S. L. 1999. *The Moon and the Western Imagination*. Tucson: University of Arizona Press.
   Fascinating story of the role of the Moon in the history of culture and science. Well written and interesting.

Sheehan, W. P., and T. A. Dobbins. 2001. *Epic Moon: A History of Lunar Exploration in the Age of the Telescope*. Richmond, VA: Willman-Bell.
   The story of the astronomers who devoted themselves to learning as much about the Moon as possible in the years before we could actually go there.

Online Resources

The advent of the Internet has made a multitude of historical and scientific documents available for reference and enlightenment. Here are a few Web sites that contain useful information that expands upon and adds to the ideas presented in this book.

Spudis Lunar Resources: http://www.spudislunarresources.com
  My personal site, containing papers, documents, graphics, and audiovisual materials supporting the ideas discussed in this book. A special section labeled on the home page ("Links") consists partly of unpublished documents that make up critical parts of the history of the Vision for Space Exploration. I also write blog posts that discuss current issues in space science and policy.

Develop Cislunar Space Next: http://www.cislunarnext.org
  A Web site that I created devoted to the development of cislunar space, including the utilization of lunar resources to create new spacefaring capabilities.

The Lunar and Planetary Institute: http://www.lpi.usra.edu
  Maintains an unparalleled resource of historical documents and lunar data. The section on the Moon contains lunar atlases, image libraries, maps, documents, and other materials related to the exploration of the Moon, past and future.

The Apollo Lunar Surface Journal: http://www.hq.nasa.gov/alsj
  Transcripts of missions operations, one of the premier documents the Apollo voyages of lunar exploration, images, videos, audios, and hundreds of other goodies. To simply call it glorious is to damn it with faint praise—exploring it merits many hours of your time.

Apollo Image Gallery: http://www.apolloarchive.com/apollo_gallery.html
  The site I always go to when I need a specific digital picture from one of the Apollo missions. Organized by mission, this collection is a wonderful asset.

History of Space Policy: http://www.hq.nasa.gov/office/pao/History/spdocs.html
  A collection of various policy papers, documents, and reports that detail national policy on civil space over time.

Lunar Reconnaissance Orbiter Camera Quickmap: http://target.lroc.asu.edu/q3
  Web-based GIS system that allows you to view any spot on the Moon from a great distance or close-up. Overlays include nomenclature, other data sets (e.g., Mini-RF radar, topography), and even current lighting conditions. A dream site for fans of the Moon.

Film and Video

Documentaries and History

Documentary films can be an enjoyable way to absorb and understand technical information about the history of the space program and lunar exploration. The following are some
of the ones that I enjoyed.

For All Mankind. 1989. National Geographic Video, 79 min.
   A documentary made up of footage from all of the Apollo missions, artistically
   combined into a single continuous narrative on how we explored the Moon.

In the Shadow of the Moon. 2007. Discovery Films, 100 min.
   The oversized role of the Apollo missions on the lives of those who
   flew them. Includes many interviews with the astronauts.

Moon Machines. 2008. Discovery Channel, 6 episodes, 60 min. each.
   Series on the major pieces of the Apollo system—the launch vehicle, the
   spacecraft, and guidance computers. The history of an engineering marvel.

To the Moon. 1999. Nova, WGBH-Boston, 120 min.
   The story of the Moon race of the 1960s. Includes interviews with all
   the principals: engineers, managers, astronauts, and scientists.

When We Left Earth: The NASA Missions. 2008. Discovery Channel, 6 episodes, 60 min. each.
   Series that compiles thousands of hours of NASA film and video
   into a narrative of humankind's first steps into the cosmos.

Lunar Feature Film Classics

Let us finally pay homage to the power of imagination. These movies are at the top of my list.

Destination Moon. 1950. Sinister Cinema Video, 91 min.
   Based on a Robert Heinlein short story, this film, produced by George Pal,
   tried to "educate" the public back at the dawn of the space age about things
   to come. A milestone science-fiction film that includes wonderful space art
   by the great Chesley Bonestell. Listen to the description of the Moon by the
   spaceship crew and compare it to the uncannily similar words of Neil Armstrong
   and Buzz Aldrin, only twenty years (but an emotional lifetime) later.

2001: A Space Odyssey. 1968. MGM Video, 139 min.
   The ultimate space movie—philosophical, intellectual, emotional, profound.
   The film, a masterpiece by Stanley Kubrick, takes great pride in getting every
   technical detail right, even down to the subtleties of weightlessness and artificial

gravity. So how come the Moon changes its phase forward, backward, and in eight-day leaps during the voyage between the space station and Clavius Base (a lousy place for a lunar outpost, by the way)? During the scene at Tycho, having the Earth appear so low on the horizon is also wrong (Tycho is at 43°S latitude, so the Earth would appear halfway between the horizon and directly overhead). Still, there's nothing like it for the "feel" of spaceflight.

Apollo 13. 1995. Universal Pictures, 130 min.
   This film, directed by Ron Howard, captures the spirit and substance of Apollo spaceflight. The space scenes are realistic (many were filmed in real "microgravity" of the NASA KC-135 aircraft) and gripping. As usual with films about the Moon (except for *Destination Moon*), many liberties are taken with lunar geography; for example, while flying over Tsiolkovsky, on the far side, crew says that they "can look up towards Mare Imbrium," which is on the opposite, near side hemisphere.

From the Earth to the Moon. 1998. HBO Films, 12 episodes, 60 min. each
   Excellent TV miniseries telling the saga of the Apollo program, from its birth to its end. A few clunker episodes (for example, the Apollo 13 episode focuses on the media and is pretty worthless—watch the film *Apollo 13* instead).

ILLUSTRATION
CREDITS

INDEX